The Immortalized Words of the Past

The Immortalized Words of the Past

prepared under the supervision of
Ralph M. Lewis, F.R.C.

Rosicrucian Library
Volume XLIV

Supreme Grand Lodge of AMORC, Inc.
Printing and Publishing Department
San Jose, California

First Edition, 1986

Library of Congress Catalogue Card No.: 85-063539
ISBN 0-912057-42-4

Printed and Bound in U.S.A.

The Rosicrucian Library

Volume

I	Rosicrucian Questions and Answers with Complete History of the Order
II	Rosicrucian Principles for the Home and Business
III	The Mystical Life of Jesus
IV	The Secret Doctrines of Jesus
V	Unto Thee I Grant (Secret Teachings of Tibet)
VI	A Thousand Years of Yesterdays (A Revelation of Reincarnation)
VII	Self Mastery and Fate with the Cycles of Life (A Vocational Guide)
VIII	Rosicrucian Manual
IX	Mystics at Prayer
X	Behold the Sign (A Book of Ancient Symbolism)
XI	Mansions of the Soul (A Cosmic Conception)
XII	Lemuria—The Lost Continent of the Pacific
XIII	The Technique of the Master
XIV	The Symbolic Prophecy of the Great Pyramid
XVI	The Technique of The Disciple
XVII	Mental Poisoning
XVIII	Glands—The Mirror of Self
XXII	The Sanctuary of Self
XXIII	Sepher Yezirah
XXV	Son of the Sun
XXVI	The Conscious Interlude
XXVII	Essays of a Modern Mystic
XXVIII	Cosmic Mission Fulfilled
XXIX	Whisperings of Self
XXX	Herbalism Through the Ages
XXXI	Egypt's Ancient Heritage
XXXII	Yesterday Has Much to Tell
XXXIII	The Eternal Fruits of Knowledge
XXXIV	Cares That Infest
XXXV	Mental Alchemy
XXXVI	Messages from the Celestial Sanctum
XXXVII	In Search of Reality
XXXVIII	Through the Mind's Eye
XXXIX	Mysticism—The Ultimate Experience
XL	The Conscience of Science and Other Essays
XLI	The Universe of Numbers
XLII	Great Women Initiates
XLIII	Increase your Power of Creative Thinking
XLIV	The Immortalized Words of the Past

(Other volumes will be added from time to time. Write for complete catalogue.)

CONTENTS

Chapter Page

Preface 9

1. Ptah-hotep (fl. 2400 B.C.) 11
2. Lao-Tse (604 B.C. - ?) 17
3. Buddha (543 B.C. - 463 B.C.) 23
4. Herodotus (484 B.C. - 430 B.C.) 29
5. Socrates (470 B.C. - 399 B.C.) 33
6. Aristotle (384 B.C. - 322 B.C.) 39
7. Epicurus (341 B.C. - 270 B.C.) 43
8. Cicero (106 B.C. - 43 B.C.) 47
9. Lucretius (98 B.C. - 55 B.C.) 53
10. Seneca, The Younger (4 B.C. - 65 A.D.) 57
11. Epictetus (50 A.D. - 130 A.D.) 61
12. Marcus Aurelius (121 - 180) 65
13. Plotinus (205 - 270) 69
14. St. Augustine (354 - 430) 75
15. Mohammed (570 - 632) 81
16. Avicenna (979 - 1037) 85
17. St. Anselm (1033 - 1109) 89
18. Moses Maimonides (1135 - 1204) 95
19. Roger Bacon (1214 - 1292) 99
20. Jan van Ruysbroeck (1293 - 1381) 103
21. St. Thomas Aquinas (1225 - 1274) 109
22. Meister Eckhart (1260 - 1328) 115
23. Thomas à Kempis (1379 - 1471) 121
24. Nicholas Copernicus (1473 - 1543) 125
25. Paracelsus (1493 - 1541) 131
26. Michel de Montaigne (1533 - 1592) 135
27. Sir Francis Bacon (1561 - 1626) 139
28. Johannes Kepler (1571 - 1630) 143
29. Jacob Boehme (1575 - 1624) 147
30. Thomas Hobbes (1588-1679) 151
31. Johann Amos Comenius (1592 - 1670) 155
32. Rene Descartes (1596 - 1650) 161
33. Thomas Vaughan (1622 - 1666) 167
34. Baruch Spinoza (1632 - 1677) 173
35. John Locke (1632 - 1704) 177

Chapter Page

36. Nicolas Malebranche (1638 - 1715) 181
37. Sir Isaac Newton (1642 - 1727) 185
38. William Penn (1644 - 1718) 191
39. Gottfried Wilhelm Leibnitz (1646 - 1716) 199
40. George Berkeley (1685 - 1753) 203
41. Benjamin Franklin (1706 - 1790) 207
42. David Hume (1711 - 1776) 213
43. Johann Gottfried von Herder (1744 - 1803) 217
44. Pierre Simon La Place (1749-1827) 221
45. Johann Gottlieb Fichte (1762 - 1814) 225
46. Sir Humphry Davy (1778 - 1829) 231
47. Arthur Schopenhauer (1788 - 1860) 235
48. Thomas Carlyle (1795 - 1881) 241
49. Auguste Comte (1798 - 1857) 245
50. Honoré de Balzac (1799 - 1850) 249
51. Ralph Waldo Emerson (1803 - 1882) 253
52. Ludwig Andreas Feuerbach (1804 - 1872) 257
53. Charles Darwin (1809 - 1882) 261
54. Sir William Crookes (1832 - 1919) 269
55. Ernst Haeckel (1834 - 1919) 273
56. John Fiske (1842 - 1901) 277
57. Elbert Hubbard (1856 - 1915) 283
58. Albert Einstein (1879 - 1955) 289

Preface

The greatest wealth that a civilization has is not alone its buried natural resources. Most of such could not contribute to the advantage of mankind without *knowledge* of its value. This knowledge is of what man has dreamed, conceived, and transmuted into reality.

The advance of mankind is determined by relation of the past to the present, and which also is often a prescience of the future. There is a touch of the divine, of *immortality,* in the preserved works of the great thinkers of the past. At least they gave us a foundation upon which to stand to envision a tomorrow.

This, then, is the purpose of this collection of writings entitled *The Immortalized Words of the Past.*

—Ralph M. Lewis, F.R.C

CHAPTER 1

PTAH-HOTEP

OF ALL THE GREAT CIVILIZATIONS of the past, Egypt put the most emphasis on immortality, life after death. This other world was to be an exalted state of existence, far transcending what this world could afford mankind. The required qualities to attain this state of afterlife were then, necessarily severe. The qualifications centered on the value of *good*, in other words, what was to be the conduct of man that would constitute this righteousness.

The mistreatment of fellow humans was deplored, in principle at least. Conversely, that which brought happiness to others, as well as to one's self, was considered an example of the ideals accepted in the next world. Simply, such conduct was thought to be of a Divine quality.

In the Ancient Mystery Schools of Egypt, such as the Osirian, and in the Pyramid texts, these virtues were defined. The earliest specific formulation of *right conduct* to be found anywhere was in Egyptian literature. *The Precepts of Ptah-hotep* was composed, it is thought, in the reign of Snefru, a vigorous and prosperous King of the IV Dynasty. We know of these works only from copies, the most complete being the papyrus which was given to the Bibleotheque Royal, in Paris, by E. Prisse d'Avennes in 1847. This document was probably written about the period of the XVIII Dynasty and may have also contained writings reflecting the current opinion of Egyptians at that latter time.

According to the introduction of the document, an appeal was made to the Sovereign King Isesi by the aged Vizier, Ptah-hotep, who, feeling the infirmity of his age, begged permission to instruct his son to be prepared for the duties of an official career, as an assistant and successor to his father. The King consented, and the old Vizier thereupon admonished his son not to misuse the wisdom he would receive, and to be humble in its acceptance and application.

It is interesting to note that the writings of Ptah-hotep (c.2400 B.C.) constitute a series of moral aphorisms, similar to those found in books of wisdom such as the apocryphal Ecclesiasticus, or the Biblical Book of Proverbs. They are pragmatic in the sense that they represent the *experience* of men of the world. It was evident that worldly success, which was important, depended upon *right conduct*. In fact, one-third of the Vizier's fifty-three paragraphs of wisdom are devoted to such admonishments. The following are examples of these ancient maxims.

RESPECT OF THE FATHER

"How worthy it is when a son hearkens to his father! If the son of a man receives what his father says, none of his projects will miscarry. Instruct as thy son one who hearkens, who shall be successful in the opinion of the princes, who directs his mouth according to that which is said to him . . . How many mishaps befall him who hearkens not! The wise man rises early to establish himself, but the fool is in trouble. As for the fool who hearkens not there is none who has done anything for him. He regards wisdom as ignorance, and what is profitable as useless. A son who hearkens . . . reaches old age, he attains reverence. He speaks likewise to his own children, renewing the instruction of his father . . . He speaks with his children, then they speak to their children."

Right Conduct

"Take when he gives to thee what he puts before thee, but do not look at what is before him, look at what is before thee, and bombard him not with many glances . . . Turn thy face downward until he addresses thee, and speak only when he has addressed thee. Laugh when he laughs, so shalt thou be very agreeable to his heart. One knows not what is in the heart."

Recognition of Authority

"Have no knowledge of his former low estate . . . be respectful towards him because of what he has achieved; for substance cometh not of itself. Be silent, for it is better than teftef-flowers. Speak thou when thou knowest that thou solvest difficulties. It is a craftsman who speaks in council and speech is more difficult than any craft. Thy food hangs upon his mood, the belly of one loved is filled, thy back shall be clothed thereby . . . Let thy mind be deep and thy speech scanty . . . Let thy mind be steadfast as long as thou speakest. May the princes who shall hear it say, 'How seemly is that which comes out of his mouth!'"

Establishing Good Will

"If thou hast become great after thou wert little, and hast gained possessions after thou wert formerly in want, . . . be not unmindful of how it was with thee before. Be not boastful of thy wealth, which has come to thee as a gift of the god. Thou art not greater than another like thee to whom the same has happened."

Friendship

"If thou searchest the character of a friend, ask no questions, approach him and deal with him when he is alone . . . Disclose his heart in conversation. If that which he has seen come forth from him, he do aught that makes thee ashamed for him, . . . do not answer."

Family Relationship

"If thou art a successful man establish to thyself a wife as the heart's mistress. Fill her body, clothe her back. The recipe for her limbs is ointment. Make her heart glad as long as thou livest. She is a profitable field for her lord."

On Fatherhood

"If thou art a successful man and establishest thy household and begettest a son having the favour of the god, if he lives correctly, inclines to thy character, hearkens to thy instruction, while his purposes are worthy in thy house, and he conserves thy possessions as should be, then seek for him every good thing. He is thy son, whom thy ka has begotten for thee. Separate not thy affection from him . . . If he errs and transgresses thy purposes, and does not observe thy instruction, his purposes are evil and he opposes all that thou sayest, his mouth is defiled with evil speech, . . . thou shalt drive him away; he is not thy son, for he has not been born to thee."

Warning About Infidelity

"If thou desirest to establish friendship in a house into which thou enterest, whether as lord, as brother, or as friend wheresoever thou enterest in, beware of approaching the women. The place where they are is not seemly, and it is not wise to intrude upon them. A thousand men are undone for the enjoyment of a brief moment like a dream. Men gain only death for knowing them. When one is intoxicated with shining limbs, then they become but herset-stone. A little, an instant, like a dream, and the end is death."

Right Exercise Of Authority

"If thou art an administrator, be gracious when thou hearest the speech of the petitioner. Do not assail him until he has cleaned out his belly of what he thought to say to thee. He who is suffering wrong desires that his heart be cheered to accomplish that on account of which he has come... It is an ornament of the heart to hear kindly. If thou art an administrator issuing ordinances for the multitude, seek for thee every excellent precedent, that thy ordinance may endure without error therein. Great is righteousness; its dispensation endures, nor has it been overthrown since the time of its maker; for punishment is inflicted on the transgressor of its laws . . . Although misfortune may carry away wealth . . . the power of righteousness is that it endures, so that a man may say 'It is a possession of my father'. Hold fast the truth and transgress it not, even though the report be not one pleasing the heart."

Good Character

"Attain character . . . make righteousness to flourish and thy children shall live. Precious to a man is the virtue of his son, and good character is a thing remembered. If thou hearkenest to this which I have said to thee, all the fashion of thee will be according to the ancestors. As for the righteousness thereof, it is their worth; the memory thereof shall not vanish from the mouths of men, because their maxims are worthy. Every word will be carried on, it shall not perish in this land forever, and it will make worthy the utterance according to which the princes speak... When a worthy reputation arises with him who is thy chief, it shall be excellent forever and all its wisdom shall be for eternity. As for the wise man, his soul rejoiceth to make his worth enduring thereby on earth. A wise man is recognized by that which he knows. His heart is the balance for his tongue, his lips are correct when he speaks, and his eyes in seeing; his ears together hear what is profitable for his son, who does righteousness and is free from lying. Established is the man whose standard is righteousness, who walketh according to its way."

Responsibility Of A Messenger

"If thou are one of the trusted ones, whom one great man sendeth to another, act rightly in the matter when he sendeth thee. Thou shalt deliver the message as he saith it. Be not secretive concerning what may be said to thee, and beware of any forgetfulness. Hold fast to the truth and overstep it not, even if thou (therewith) recountest nothing that is gratifying. Beware of worsening words, such as might make one great man contemptible to the other through the manner of speech of all men. 'A great man, an insignificant one'—that is what the ka abhorreth."

Covetousness

"If thou desirest thy conduct to be good, to set thyself free from all that is evil, then beware of covetousness, which is a malady, diseaseful, incurable. Intimacy with it is impossible; it maketh the sweet friend bitter, it alienateth the trusted one from the master, it maketh bad both father and mother, together with the brothers of the mother, and it divorceth a man's wife. It is a bundle of every kind of evil, and a bag of everything that is blameworthy. Long lived is the man whose rule of conduct is right, and who goeth in accordance with his course; he winneth wealth thereby, but the covetous hath no tomb."

Ptah-hotep retired as Grand Vizier (Minister of State) under the Pharaoh Isesi of the V Dynasty of the Old Kingdom of Egypt. He concluded his collection of wise maxims with these words: "I have attained one hundred and ten years of life, while the King gave me rewards—because I did righteousness for the King even unto the grave."

CHAPTER 2

LAO-TSE

LAO-TSE, AN ANCIENT CHINESE sage, was the founder of the philosophy known as Taoism, the most widely spead "popular" religion of China. There are few facts concerning him that are reliable, for his life has been so surrounded with legends invented by later Taoism that the truth has been somewhat obscured. One of these legends stated that his mother carried him in her womb for seventy years and that he was born with white hair, thus causing him to be named Lao-Tse, which can mean either "old boy" or "venerable philosopher." His birth is placed at 604 B.C., for it is well-established that he was a contemporary of Confucius. He became librarian and archivist of the Chou dynasty and is supposed to have had some influence on Confucius, who visited him about 517 B.C. According to the Chinese historian, Sze-ma Ch'ien, he resided at the capital of Chow, but the evils of the world and the decaying dynasty caused him to sever his worldly connections, but not before he complied with the request of the last civilized man he met, the warden of the outer gate, to put in book form the essence of his great teachings—the doctrine of "the Way."

Lao-Tse attempted to make no converts to the Tao, which indicates that he may have realized that he was far ahead of his age and the people were not ready to receive the Tao teachings. Being of a retiring nature and always seeking obscurity for himself, it was in character for him to withdraw himself from the world. His age was reputed to be from 160 to 200 years; and the time or place of his transition unknown.

Lao-Tse professes to be merely a transmitter of earlier knowledge. In fact, he asserts that his teachings are really those of Hwang-Ti, the legendary civilizer of China, c. 2704 B.C., whose teachings in turn are supposed to have been traced back at least five thousand years previous. The *Tao-Teh King*, the gospel of Taoism, is the only surviving work of Lao-Tse and is ranked with the greatest books of the world.

TAO-TEH KING OR BOOK OF THE VALUES OF THE TAO

VIII

Transcendent Goodness is like water.

Water is peaceful and extends its beneficent action throughout Nature, not even disdaining those gloomy depths which the vulgar look upon with horror, for water works much as the Tao does.

Now, the term "Goodness" has a variety of applications. It may refer to the quality of the ground upon which a house stands; or to profundity in a thinker; or to sincerity in a speaker; or to well-ordered government; or to a capacity for doing; or to punctuality; but it is only when goodness is used in reference to freedom from contention that it can be considered faultless.

IX

It is easier to carry an empty vessel than a full one.

The point which is often felt after it has been sharpened will soon become blunt.

The hall which is filled with silver and gold will not long retain its contents.

He who bears wealth and honor arrogantly will work out his own destruction.

When meritorious services have led to fame, it is time to follow the heavenly rule and retire into obscurity.

X

He who makes the investigation of his spiritual nature his chief object will be able to bring all his studies to a focus, and this concentration of his energies will render him capable of arriving at a condition of sensibility to impressions similar to that which belongs to a young child.

He who is able to wash himself clean from all obscure and gloomy thoughts will become sound in mind, and—should he be a ruler—if he govern his people on principles founded on love, he will be able to remain in perfect repose and peace as he watches the processes of Nature proceed around him. He will be as the brood hen who carries on her work when in a state of perfect rest; and who, whilst the light of intelligence may overspread the world, is able, without knowledge, to procreate and nourish; to bring forth, and not retain possession; to increase and multiply, and not to hold in subjection; to act, and not to depend upon others for assistance.

Well indeed may this be called a deep and impenetrable mystery.

XI

The thirty spokes of a chariot-wheel and the nave to which they are attached would be useless, but for the hollow space in which the axle turns.

The vase molded out of clay would be useless, but for the empty space left for its contents.

The door and window frames of a house would be useless, but for the empty spaces they enclose, which permit of ingress and egress, and the admission of light and air.

This teaches us that, however beneficial the *material* may be to us, without the *immaterial* it would be useless.

XII

The eye is dazzled by a variety of colors,

The ear is deafened by a diversity of sounds,

The taste is vitiated by a mixture of flavors,

The mind is excited by excessive exercise,

And the character is ruined by seeking to be rich.

Hence it is that the wise man prefers to be emotional rather than to be sensuous, and it is through this that his perceptive faculties become cultivated, so that he is able to arrive at just conclusions.

XIII

There are two sayings which require explanation—

"Promotion and degradation alike give rise to fear," and "Suffering and honor are alike corporeal."

The meaning of the first is, that he who has been promoted lives in fear that he may be degraded, whilst he who has been degraded is haunted by the dread that his degradation may be continued.

With respect to the second saying, it means that the sense of suffering is a consequence of corporeal existence; without a body there could be no bodily pain, and for the same reason there would be no personality on whom honor could be bestowed.

This is why he who does honor to his own person, or he who bestows the same love upon others as he does upon himself, may be entrusted with the government of an empire.

XIV

That which you look at and can not see is called "invisible."

That which you listen to and can not hear is called "inaudible."

That which you seize upon and can not grasp is called "intangible."

These three definitions are difficult of realization when taken singly, let us therefore try what can be done by bringing them together and uniting them in One.

The three negations now form a single combination, but if we scrutinize it closely, no matter in what aspect we may regard it, we shall find nothing either hidden or revealed; and let us be careful not to define it or give it a name, or it will escape from us and become even more subtle than it was before. This is what is meant by "seeking to define the indefinable," and "to establish a resemblance between things which have no real existence."

The Tao is indeed a deep mystery. We can not recognize its presence; if we advance toward the Tao we can not see what is behind it; if we follow the Tao we can not see what is before it. Yet, if we would gain a knowledge of our present lives, we must hold on to the Tao of the Past, and the only clue which will lead us up to it is a knowledge of the processes which formed the beginning of that Past.

XV

The virtues of the olden time, as practiced by the Sages, come down to us in such an exiguous, indefinite, and obscure form that it is very difficult for us to understand them. I will, however, do my best to make them clear.

That which the Sages took a pleasure in doing may be likened to the wading across a swollen torrent in mid-winter.

Their caution resembled that which is produced by a fear of our associates and of those who live in our neighborhood.

Their carriage was as the bearing of a guest toward his host.

Their self-effacement was as the melting away of an icicle.

In their indignation they were rough as a piece of unplaned wood.

Their influence was as far-reaching as the flow of a mountain torrent, and like the torrent it became turbid through its own movement.

Now who is there capable of cleansing the impurities of his nature by tranquillity and rest? And who is there capable of producing a state of perfect repose by the long-continued calm of a peaceful life?

In conclusion: Those who affect to cherish these principles, and yet have no desire to carry them out in their entirety, will become capable of committing vile actions, and so remain to the end of their lives in an unreformed and imperfect condition.

XXVII

A good walker moves lightly over the ground, and his footsteps leave no trace.

A good speaker is accurate and keeps his temper.

A good reckoner needs no tablets.

A good smith needs no wooden bars, yet the doors he fastens can not be opened.

A skilful joiner needs no cords to keep his work together.

In the same manner, it is through the skill and ability of the Sage that his fellow men are aided without one of them being discarded or lost, and it is the same when he deals with the brute creation or material objects.

This is what is called being "doubly enlightened," and hence it is that the skilful man becomes the unskilful man's master, and the unskilful man becomes the skilful man's slave.

When the slave does not honor his master, and the master does not love his slave, although they may both have a knowledge of what is suitable, they will be guilty of gross stupidity.

This may be considered an abstract of the leading principles belonging to a very difficult and subtle subject.

XXXII

"The Tao is unchanging and has no name."

Now, although this statement is so short and so simple, the world can not take it in. Yet if kings and princes were but to receive it, there is nothing under Heaven which would not resort to them, and it would produce a spirit of harmony which would descend upon the Empire like a fragrant dew, so that the people would no longer require to receive orders from their superiors, but would be rendered capable of controlling their own actions.

But when a name was given to the Great First Cause,* which has been continued to this day, the knowledge I speak of became arrested, and we soon ceased to be familiar with that which it withheld from us.

Ah! if the right knowledge of the Tao were but spread through the Empire, it would become like the ocean and great rivers into which the rivulets and streams continuously flow.

*In this poetic version, the original translator had substituted the word *God* for the word *Tao*. The phrase *Great First Cause* is not found in the original Chinese version.

CHAPTER 3

BUDDHA

SIDDHARTHA GAUTAMA IS KNOWN as Buddha, the Enlightened One. According to various traditions and some reliable historical records, the exact birthdate of Buddha is not known, but is generally fixed as 543 B.C., by Ceylonese reckoning. Nearly all of the records of his life agree, however, that he was the son of a king of the Sakyas, a people of the warrior caste near the Himalayas, and north of the Kosala Kingdom, who claimed to belong to the Gautama Clan. Hence Buddha was called Gautama Buddha. It appears that it had been prophesied at his conception that occasions would arise in his life which would tempt him to renounce the world if he ever came in contact with the sick, the old, and the dead. To prevent such a break in the line of the family power, the child was raised in great luxury where he saw nothing but that which was of the very finest, the most elevating, and cultural. Apparently this did not prevent him from thinking, however, and in his meditations his soul was led toward contacts with humanity, and at the age of twenty-nine he left home, wandering for six years seeking the spiritual enlightenment which his soul craved. It came to him while meditating under a tree known as the BO-TREE, which is always associated with his life. From this enlightenment he derived the complete title of his name. After this incident he formed an order of monks to which, later in his lifetime, an order of nuns was added, and he spent the remainder of his life wandering and preaching the newly revealed doctrines, not only to the monks who were members of his order, but to the laity generally. His transition occurred at the age of eighty.

THE DHARMAPADA

This is the Dharmapada, the path of religion pursued by those who are followers of the Buddha:

Creatures from mind their character derive; mind-marshalled are they, mind-made. Mind is the source either of bliss or of corruption.

By oneself evil is done; by oneself one suffers; by oneself evil is left undone; by oneself one is purified. Purity and impurity belong to oneself, no one can purify another.

You yourself must make an effort. The Tathagatas are only preachers. The thoughtful who enter the way are freed from the bondage of Mara.

He who does not rouse himself when it is time to rise; who, though young and strong, is full of sloth; whose will and thoughts are weak; that lazy and idle man will never find the way to enlightenment.

If a man hold himself dear, let him watch himself carefully; the truth guards him who guards himself.

If a man makes himself as he teaches others to be, then, being himself subdued, he may subdue others; one's own self is indeed difficult to subdue.

If some men conquer in battle a thousand times a thousand men, and if another conquer himself, he is the greatest of conquerors.

It is the habit of fools, be they laymen or members of the clergy, to think, "this is done by *me*. May others be subject to *me*. In this or that transaction a prominent part should be played by *me*." Fools do not care for the duty to be performed or the aim to be reached, but think of their self alone. Everything is but a pedestal of their vanity.

Bad deeds, and deeds hurtful to ourselves, are easy to do; what is beneficial and good, that is very difficult.

If anything is to be done, let a man do it, let him attack it vigorously!

Before long, alas! this body will lie on the earth, despised, without understanding, like a useless log; yet our thoughts will endure. They will be thought again, and will produce action. Good thoughts will produce good actions, and bad thoughts will produce bad actions.

Earnestness is the path of immortality, thoughtlessness the path of death. Those who are in earnest do not die; those who are thoughtless are as if dead already.

Those who imagine they find truth in untruth, and see untruth in truth, will never arrive at truth, but follow vain desires. They who know truth in truth, and untruth in untruth, arrive at truth, and follow true desires.

As rain breaks through an ill-thatched house, passion will break through an unreflecting mind. As rain does not break through a well-thatched house, passion will not break through a well-reflecting mind.

Well-makers lead the water wherever they like; fletchers bend the arrow; carpenters bend a log of wood; wise people fashion themselves; wise people falter not amidst blame and praise. Having listened to the law, they become serene, like a deep, smooth, and still lake.

If a man speaks or acts with an evil thought, pain follows him as the wheel follows the foot of the ox that draws the carriage.

An evil deed is better left undone, for a man will repent of it afterwards; a good deed is better done, for having done it one will not repent.

If a man commits a wrong let him not do it again; let him not delight in wrongdoing; pain is the outcome of evil. If a man does what is good, let him do it again; let him delight in it; happiness is the outcome of good.

Let no man think lightly of evil, saying in his heart, "It will not come nigh unto me." As by the falling of water-drops a water-pot is filled, so the fool becomes full of evil, though he gather it little by little.

Let no man think lightly of good, saying in his heart, "It will not come nigh unto me." As by the falling of water-drops a water-pot is filled, so the wise man becomes full of good, though he gather it little by little.

He who lives for pleasure only, his senses uncontrolled, immoderate in his food, idle, and weak, him Mara, the tempter, will certainly overthrow, as the wind throws down a weak tree. He who lives without looking for pleasures, his senses well-controlled, moderate in his food, faithful and strong, him Mara will certainly not overthrow, any more than the wind throws down a rocky mountain.

The fool who knows his foolishness, is wise at least so far. But a fool who thinks himself wise, he is a fool indeed.

To the evil-doer wrong appears sweet as honey; he looks upon it as pleasant so long as it bears no fruit; but when its fruit ripens, then he looks upon it as wrong. And so the good man looks upon the goodness of the Dharma as a burden and an evil so long as it bears no fruit; but when its fruit ripens then he sees its goodness.

A hater may do great harm to a hater, or an enemy to an enemy; but a wrongly-directed mind will do greater mischief unto itself. A mother, a father, or any other relative will do much good; but a well-directed mind will do greater service unto itself.

He whose wickedness is very great brings himself down to that state where his enemy wishes him to be. He himself is his greatest enemy. Thus a creeper destroys the life of a tree on which it finds support.

Do not direct thy thought to what gives pleasure, that thou mayest not cry out when burning, "This is pain." The wicked man burns by his own deeds, as if burnt by fire.

Pleasures destroy the foolish; the foolish man by his thirst for pleasures destroys himself as if he were his own enemy. The fields are damaged by hurricanes and weeds; mankind is damaged by passion, by hatred, by vanity, and by lust.

Let no man ever take into consideration whether a thing is pleasant or unpleasant. The love of pleasure begets grief and the dread of pain causes fear; he who is free from the love of pleasure and the dread of pain knows neither grief nor fear.

He who gives himself to vanity, and does not give himself to meditation, forgetting the real aim of life and grasping at pleasure, will in time envy him who has exerted himself in meditation.

The fault of others is easily perceived, but that of oneself is difficult to perceive. A man winnows his neighbor's faults like chaff, but his own fault he hides, as a cheat hides the false die from the gambler.

If a man looks after the faults of others, and is always inclined to take offense, his own passions will grow, and he is far from the destruction of passions.

Not about the perversities of others, not about their sins of commission or omission, but about his own misdeeds and negligences alone should a sage be worried.

Good people shine from afar, like the snowy mountains; bad people are concealed, like arrows shot by night.

If a man by causing pain to others, wishes to obtain pleasure for himself, he, entangled in the bonds of selfishness, will never be free from hatred.

Let a man overcome anger by love, let him overcome evil by good; let him overcome the greedy by liberality, the liar by truth!

For hatred does not cease by hatred at any time; hatred ceases by nonhatred, this is an old rule.

Speak the truth, do not yield to anger; give, if thou art asked; by these three steps thou wilt become divine.

Let a wise man blow off the impurities of his self, as a smith blows off the impurities of silver, one by one, little by little, and from time to time.

Lead others, not by violence, but by righteousness and equity.

He who possesses virtue and intelligence, who is just, speaks the truth, and does what is his own business, him the world will hold dear.

As the bee collects nectar and departs without injuring the flower, or its color or scent, so let a sage dwell in the community.

If a traveler does not meet with one who is his better, or his equal, let him firmly keep to his solitary journey; there is no companionship with fools.

Long is the night to him who is awake; long is a mile to him who is tired; long is life to the foolish who do not know the true religion.

Better than living a hundred years, not seeing the highest truth, is one day in the life of a man who sees the highest truth.

Some form their Dharma arbitrarily and fabricate it artificially; they advance complex speculations and imagine that good results are attainable only by the acceptance of their theories; yet the truth is but one; there are not different truths in the world. Having reflected on the various theories, we have gone into the yoke with him who has shaken off all sin. But shall we be able to proceed together with him?

The best of ways is the eightfold path. This is the path. There is no other that leads to the purifying of intelligence. Go on this path! Everything else is the deceit of Mara, the tempter. If you go on this path, you will make an end of pain! Says the Tathagata, the path was preached by me, when I had understood the removal of the thorn in the flesh.

Not only by discipline and vows, not only by much learning, do I earn the happiness of release which no worldling can know. Bhikshu, be not confident as long as thou hast not attained the extinction of thirst. The extinction of evil desire is the highest religion.

The gift of religion exceeds all gifts; the sweetness of religion exceeds all sweetness; the delight in religion exceeds all delights; the extinction of thirst overcomes all pain.

Few are there among men who cross the river and reach the goal. The great multitudes are running up and down the shore; but there is no suffering for him who has finished his journey.

As the lily will grow full of sweet perfume and delight upon a heap of rubbish, thus the disciple of the truly enlightened Buddha shines forth by his wisdom among those who are like rubbish, among the people that walk in darkness.

Let us live happily then, not hating those who hate us! Among men who hate us let us dwell free from hatred!

Let us live happily then, free from all ailments among the ailing! Among men who are ailing let us dwell free from ailments!

Let us live happily then, free from greed among the greedy! Among men who are greedy let us dwell free from greed!

The sun is bright by day, the moon shines by night, the warrior is bright in his armor, thinkers are bright in their meditation; but among all the brightest with splendor day and night is the Buddha, the Awakened, the Holy, Blessed.

CHAPTER 4

HERODOTUS

HERODOTUS IS CALLED THE father of history. He is the first eminent historian. He recorded actual facts and incidents rather than the imaginings of man's mind or romances about the lives and conduct of the gods. He was born at Halicarnassus about 484 B.C., in Asia Minor. The city in which he was born was at that time Persian territory; therefore, although he was Greek, he was a Persian subject and continued as a Persian subject until thirty or thirty-five years of age when he started his travels. He visited nearly all the parts of the civilized world at that time which fringed the shores of his native country. It is estimated that he traveled a total of about seventeen hundred miles, which, considering the ancient mode of travel, is not as insignificant as it might at first seem.

His close observation of the habits and customs of the people and countries that he visited, and which he recorded, won the approval of the people at Athens. So great was their admiration for him that he was voted a sum of ten talents ($12,000). He was severely criticized by ancient writers, being accused of untruthfulness. He was subject to the usual attacks upon those who are famous and progressive. Modern writers acquit him of the charge of untruthfulness. He is even considered today, in the light of our modern archaeological research, to be an authority on the ancient Persian wars inasmuch as archaeological findings have supported his accounts.

Below are given excerpts from one of his works entitled, *First Hand Observations.* One cannot help but admire his descriptive powers. At the same time it permits one to know of the customs and superstitious practices of the masses in Egypt at that time. Amongst all that superstition and ignorance, the few were bound together who dared to seek knowledge and cross the threshold of the arcane mystery schools of learning.

FIRST HAND OBSERVATIONS

Regarding Egypt itself I shall extend my remarks to a great length, because there is no country that possesses so many wonders, nor any that has such a number of works which defy description. Not only is the climate different from that of the rest of the world, and the rivers unlike any other rivers, but the people also, in most of their manners and customs, exactly reverse the common practice of mankind. The women attend the markets and trade, while the men sit at home at the loom; and here, while the rest of the world works the woof up the warp, the Egyptians work it down; the women likewise carry burdens upon their shoulders, while the men carry them upon their heads. They eat their food out of doors in the streets, but retire for private purposes to their houses, giving as a reason that what is unseemly, but necessary, ought to be done in secret, but what has nothing unseemly about it, should be done openly. A woman cannot serve the priestly office, either for god or goddess, but men are priests to both; sons need not support their parents unless they choose, but daughters must, whether they choose or no.

In other countries the priests have long hair, in Egypt their heads are shaven; elsewhere it is customary, in mourning, for near relations to cut their hair close; the Egyptians, who wear no hair at any other time, when they lose a relative, let their beards and the hair of their heads grow long. All other men pass their lives separate from animals, the Egyptians have animals always living with them; others make barley and wheat their food, it is a disgrace to do so in Egypt, where the grain they live on is spelt, which some call *zea*. Dough they knead with their feet; but they mix mud, and even take up dirt, with their hands. They are the only people in the world—they at least, and such as have learnt the practice from them—who use circumcision. Their men wear two garments apiece, their women but one. They put on the rings and fasten the ropes to sails inside, others put them outside. When they write or calculate, instead of going, like the Greeks, from left to right, they move their hand from right to left; and they insist, notwithstanding, that it is they who go to the right, and the Greeks who go to the left. They have two quite different kinds of writing, one of which is called sacred, the other common.

They are religious to excess, far beyond any other race of men, and use the following ceremonies:—They drink out of brazen cups, which they scour every day: there is no exception to this practice. They wear linen garments, which they are specially careful to have always fresh washed. They practice circumcision for the sake of cleanliness, considering it better to be cleanly than comely. The priests shave their whole body every other day, that no lice or other impure thing may adhere to them when they are engaged in the service of the gods. Their dress is entirely of linen, and their shoes of the papyrus plant; it is not lawful for them to wear either dress or shoes of any other material. They bathe twice every day in cold water, and twice each night. Besides which they observe, so to speak, thousands of ceremonies. They enjoy, however, not a few advantages. They consume none of their own property, and are at no expense for anything; but every day bread is baked for them of the sacred corn, and a plentiful supply of beef and of goose's flesh is assigned to each, and also a portion of wine made from the grape. Fish they are not allowed to eat; and beans,—which none of the Egyptians ever sow, or eat, if they come up of their own accord, either raw or boiled—the priests will not even endure to look on, since they consider it an unclean kind of pulse. Instead of a single priest, each god has the attendance of a college, at the head of which is a chief priest; when one of these dies, his son is appointed in his room.

Male kine are reckoned to belong to Epaphus, and are therefore tested in the following manner:—One of the priests appointed for the purpose searches to see if there is a single black hair on the whole body, since in that case the beast is unclean. He examines him all over, standing on his legs, and again laid upon his back; after which he takes the tongue out of his mouth, to see if it be clean in respect to the prescribed marks; he also inspects the hairs of the tail, to observe if they grow naturally. If the animal is pronounced clean in all these various points, the priest marks him by twisting a piece of papyrus round his horns, and attaching thereto some sealing-clay, which he then stamps with his own signet ring. After this the beast is led away; and it is forbidden, under the penalty of death, to sacrifice an animal which has not been marked in this way.

The following is their manner of sacrifice:—They lead the victim, marked with their signet, to the altar where they are about to offer it, and setting the wood alight, pour a libation of wine upon the altar in front of the victim, and at the same time invoke the god. Then they slay the animal, and cutting off his head, proceed to flay the body. Next they take the head, and heaping imprecations on it, if there is a market-place and a body of Greek traders in the city, they carry it there and sell it instantly; if, however, there are no Greeks among them, they throw the head into the river. The imprecation is to this effect:—They pray that if any evil is impending either over those who sacrifice, or over universal Egypt, it may be made to fall upon that head. These practices, the imprecations upon the heads, and the libations of wine, prevail all over Egypt, and extend to victims of all sorts; and hence the Egyptians will never eat the head of any animal.

The disembowelling and burning are however different in different sacrifices. I will mention the mode in use with respect to the goddess whom they regard as the greatest, and honour with the chiefest festival. When they have flayed their steer they pray, and when their prayer is ended they take the paunch of the animal out entire, leaving the intestines and the fat inside the body; they then cut off the legs, the end of the loins, the shoulders, and the neck; and having so done, they fill the body of the steer with clean bread, honey, raisins, figs, frankincense, myrrh, and other aromatics. Thus filled, they burn the body, pouring over it great quantities of oil. Before offering the sacrifice they fast, and while the bodies of the victims are being consumed they beat themselves. Afterwards, when they have concluded this part of the ceremony, they have the other parts of the victim served up to them for a repast.

The male kine, therefore, if clean, and the male calves, are used for sacrifice by the Egyptians universally; but the female they are not allowed to sacrifice, since they are sacred to Isis. The statue of this goddess has the form of a woman but with horns like a cow, resembling thus the Greek representations of Io; and the Egyptians, one and all, venerate cows much more highly than any other animal. This is the reason why no native of Egypt, whether man or woman, will give a Greek a kiss, or use the knife of a Greek, or his spit, or his cauldron, or taste the flesh of an ox, known to be pure, if it has been cut with a Greek knife. When kine die, the following is the manner of their sepulture:—The females are thrown into the river; the males are buried in the suburbs of the towns, with one or both their horns appearing above the surface of the ground to mark the place. When the bodies are decayed, a boat comes, at an appointed time, from the island called Prosopitis,—which is a portion of the Delta, nine schoenes in circumference,—and calls at the several cities in turn to collect the bones of the oxen. Prosopitis is a district containing several cities; the name of that from which the boats come is Atarbechis. Venus has a temple there of much sanctity. Great numbers of men go forth from this city and proceed to other towns, where they dig up the bones, which they take away with them and bury together in one place. The same practice prevails with respect to the interment of all other cattle—the law so determining; they do not slaughter any of them.

CHAPTER 5

SOCRATES

SOCRATES WAS BORN ABOUT 470 B.C. at Athens. He is reputed to have been most ugly in his physical appearance, yet having a very congenial disposition. He was exceptionally democratic and was to be found mostly around the market places and in the public squares, delighting in interrogating those personages who particularly thought they were possessed of great knowledge or wisdom, and by his method of inquiry causing them to realize their lack of understanding. Legends say that Socrates once visited the oracle at the sanctuary of Apollo at Delphi, and was told that he was the wisest of all men. Not believing this, he set out to prove that the oracle was wrong by searching for one who could easily confound him, but never succeeded. His school or movement began the first ethical period, the search for knowledge of self, of what is right and wrong, and what constitutes the soul. He was little concerned with the cosmological theories of his predecessors. He was not an individualist. The highest good in life was to be found in serving the State and in submerging the individuality in the interests of the State. All we know of the Socratic philosophy comes to us through the dialogues of his student, Plato. Whether the words are of Socrates or put in his mouth by Plato will perhaps never be known.

We bring to you an excerpt from Plato's dialogue, the *Phaedo*, which concerns itself with Socrates' discourse upon the immortality of the soul upon the occasion of his imprisonment while awaiting his execution.

PHAEDO

Must we not, said Socrates, ask ourselves some question of this sort?—What is that which, as we imagine, is liable to be scattered away, and about which we fear? and what again is that about which we have no fear? And then we may proceed to inquire whether that which suffers dispersion is or is not of the nature of the soul—our hopes and fears as to our own souls will turn upon that.

That is true, he said.

Now the compound or composite may be supposed to be naturally capable of being dissolved in like manner as of being compounded; but that which is uncompounded, and that only, must be, if anything is, indissoluble.

Yes; that is what I should imagine, said Cebes.

And the uncompounded may be assumed to be the same and unchanging, whereas the compound is always changing and never the same?

That I also think, he said.

Then now let us return to the previous discussion. Is that idea or essence, which in the dialectical process we define as essence or true existence—whether essence of equality, beauty, or anything else—are these essences, I say, liable at times to some degree of change? or are they each of them always what they are, having the same simple self-existent and unchanging forms, and not admitting of variation at all, or in any way, or at any time.

They must be always the same, Socrates, replied Cebes.

And what would you say of the many beautiful—whether men or horses or garments or any other things which may be called equal or beautiful—are they all unchanging and the same always, or quite the reverse? May they not rather be described as almost always changing and hardly ever the same, either with themselves or with one another?

The latter, replied Cebes; they are always in a state of change.

And these you can touch and see and perceive with the senses, but the unchanging things you can only perceive with the mind—they are invisible and are not seen?

That is very true, he said.

Well then, he added, let us suppose that there are two sorts of existences—one seen, the other unseen.

Let us suppose them.

The seen is the changing, and the unseen is the unchanging?

That may be also supposed.

And, further, is not one part of us body, and the rest of us soul?

To be sure.

And to which class may we say that the body is more alike and akin?

Clearly to the seen: No one can doubt that.

And is the soul seen or not seen?

Not by man, Socrates.

And by 'seen' and 'not seen' is meant by us that which is or is not visible to the eye of man?

Yes, to the eye of man.

And what do we say of the soul?—is that seen or not seen?

Not seen.

Unseen then?

Yes.

Then the soul is more like to the unseen, and the body to the seen?

That is most certain, Socrates.

And were we not saying long ago that the soul when using the body as an instrument of perception, that is to say, when using the sense of sight or hearing or some other sense (or the meaning of perceiving through the body is perceiving through the senses)—were we not saying that the soul too is then dragged by the body into the region of the changeable, and wanders and is confused; the world spins round her, and she is like a drunkard when under their influence?

Very true.

But when returning into herself she reflects; then she passes into the realm of purity, and eternity, and immortality, and unchangeableness, which are her kindred, and with them she ever lives, when she is by herself and is not let or hindered; then she ceases from her erring ways, and being in communion with the unchanging is unchanging. And this state of the soul is called wisdom?

That is well and truly said, Socrates, he replied.

And to which class is the soul more nearly alike and akin, as far as may be inferred from this argument, as well as from the preceding one?

I think, Socrates, that, in the opinion of every one who follows the argument, the soul will be infinitely more like the unchangeable—even the most stupid person will not deny that.

And the body is more like the changing?

Yes.

Yet once more consider the matter in this light: When the soul and the body are united, then nature orders the soul to rule and govern, and the body to obey and serve. Now which of these two functions is akin to the divine? and which to the mortal? Does not the divine appear to you to be that which naturally orders and rules, and the mortal that which is subjected and servant?

True.

And which does the soul resemble?

The soul resembles the divine, and the body the mortal—there can be no doubt of that, Socrates.

Then reflect, Cebes: is not the conclusion of the whole matter this—that the soul is in the very likeness of the divine, and immortal, and intelligible, and uniform, and indissoluble, and unchangeable. Can this, my dear Cebes, be denied?

No indeed.

But if this is true, then is not the body liable to speedy dissolution? and is not the soul almost or altogether indissoluble?

Certainly.

And do you further observe, that after a man is dead, the body, which is the visible part of man, and has a visible framework, which is called a corpse, and which would naturally be dissolved and decomposed and dissipated, is not dissolved or decomposed at once, but may remain for a good while, if the constitution be sound at the time of death, and the season of the year favorable? For the body when shrunk and embalmed, as is the custom in Egypt, may remain almost entire through infinite ages; and even in decay, still there are some portions, such as the bones and ligaments, which are practically indestructible. You allow that?

Yes.

And are we to suppose that the soul, which is invisible, in passing to the true Hades, which like her is invisible, and pure, and noble, and on her way to the good and wise God, whither, if God will, my soul is also soon to go—that the soul, I repeat, if this be her nature and origin, is blown away and perishes immediately on quitting the body, as the many say? That can never be, my dear Simmias and Cebes. The truth rather is, that the soul which is pure at departing draws after her no bodily taint, having never voluntarily had connection with the body, which she is ever avoiding, herself gathered into herself; for such abstraction has been the study of her life. And what does this mean but that she has been a true disciple of philosophy, and has practiced how to die easily? And is not philosophy the practice of death?

Certainly.

That soul, I say, herself invisible, departs to the invisible world—to the divine and immortal and rational: thither arriving, she lives in bliss and is released from the error and folly of men, their fears and wild passions and all other human ills, and for ever dwells, as they say of the initiated, in company with the gods? Is not this true, Cebes?

Yes, said Cebes, beyond a doubt.

CHAPTER 6

ARISTOTLE

ARISTOTLE WAS BORN AT Stagirus, a Greek colony in Macedonia, in 384 B.C. At the age of seventeen he was sent to Athens to complete his education, and there entered the academy under Plato.

After Plato's death he left Athens and within a few years became the tutor of young Alexander—known to history as "the Great." Here he acquired great influence and, after the conquest of Persia, received 800 talents in gold (about $1,000,000) from his former pupil. In addition to this gift, Alexander sent him specimens of all curious animals and plants which were discovered on his expeditions.

When he was about fifty years old, Aristotle returned to Athens—bringing with him these vast scientific collections—and, in the Lyceum, established what has often been styled the "Peripatetic School" of philosophy. This name was descriptive of his habit of discoursing to pupils while walking about the shady paths (peripatoi) surrounding the Lyceum.

Twelve years later the Athenians accused him of being a partisan of the Macedonian dynasty—although his friendly relations with Alexander had been broken off—and he was forced to flee to Chalcis.

It is said that since Plato had found universal truth only in connection with classes of things, Aristotle started with an examination of the relation of the particular to the general, studying interrelations of things and listing them in the categories. In addition to his metaphysical works on the nature of Being, he made important additions to many departments of natural science, and although some of his works are undoubtedly lost, the remainder exceed in bulk those of any other classic Greek author. The excerpt below is from *The Nicomachean Ethics*.

THE HIGHEST HUMAN GOOD

And here we will close this digression, and return to the question of what is that highest human good of which we are in quest. It is clear that every course of action and every art has its own peculiar good; for the good sought by medicine is one, and the good sought by tactics is another; and of all other arts the same rule holds. What, then, is in each case the chief good? Surely it will be that to which all else that is done is but a means. And this in medicine will be health, and in tactics victory, and in architecture a house, and so forth in other cases; and in all free action, that is to say in all purpose or conscious choice of means to a desired end, it will be that end; for it is with this in view that we always take all the other steps in the particular action. And so, if there be but one end of all things that we do, this will be, in all human action, the chief good; while, if there be more than one, it will be their sum. Our argument, therefore, has now returned to the question from which it originally digressed, and which we must endeavour yet more thoroughly to clear up. Now, since there are clearly many and diverse ends, some of which we occasionally choose as means, such as wealth, or pipes, or instruments generally, it is evident that all of these various ends cannot be final; whereas the chief good is clearly a something absolutely final. So that, if there be but one thing alone that is final, this will be the good of which we are in quest; and, if there be more than one, then it will be the most final among them. Now we call that which is pursued for its own sake more final than that which is pursued as a means to something further; and that which is never chosen as a means we call more final than any such things as are chosen both as ends in themselves and as means to this; while, to sum up, we call that alone absolutely final which is in all cases to be chosen as an end, and never as a means. And happiness would seem to be pre-eminently such; for happiness we always choose as an end, and never as a means; while honour, and pleasure, and reason, and, generally, every kind of virtue we do indeed choose as ends (for we should choose each one of them, even if they bore no good fruit), but we choose them also for the sake of happiness, thinking that by their means we shall be happy. But happiness itself no man ever chooses for the sake of these things, or indeed as a means to aught beyond itself. And the all-sufficiency of happiness clearly leads to the same conclusion; for the final human good is always held to be all-sufficient. Nor do we understand that the range of this all-sufficient is to be restricted to the individual in a life of isolation but rather hold that it also includes his parents, and his children, and his wife, and indeed his friends generally, and his fellow-citizens, since man's true nature is to be citizen of a free state. And yet some limit must be fixed herein; for were one so to extend this as to take in a man's ancestors, and his descendants, and the friends of his friends, the circle would become infinite. This question, however, we will consider at some other time, and for the present will define as all-sufficient that which alone and by itself can make our life desirable, and supply all our needs. And we are of opinion that happiness is such. And, moreover, happiness is the most desirable of all things, in that there is nothing else which is on a par with it, and so capable of being added to it. Were not this so, then the addition of any other good, no matter how small, would evidently render it more desirable. For such addition would constitute a surplus of good; and of any two goods the greater is always the more choiceworthy. Happiness, then, is clearly a something complete in itself, and all-sufficient, forming the one end of all things done by man.

But still to say nothing more about happiness than that it is the greatest of all goods is clearly but little better than a truism, and one seems to yearn for a yet more exact and definite account. This we shall most probably obtain from the consideration of what it is that man, as man, has to do. For, as in the case of flute players, and of sculptors, and of all craftsmen, and indeed of all those who have any work of their own to do, or who can originate any special train of action, it is in this their especial work or function that their chief good and greatest welfare lie, so too ought it to be in the case of man as man, if as man he has any special functions of his own. Are we then to believe that man as carpenter, or that man as cobbler, has a function of his own, and so can originate an especial course of action; while as man he lacks this, and has no task assigned him by nature? Shall we not rather say that exactly as the eye, and the hand, and the foot, and each of the various members, evidently has its office, so too, beyond and beside all these, must be assigned an office to man, as such? And, if so, what are we to say that this office is? Life he has in common even with plants, whereas what we seek is that, whatever it is, that is especial and peculiar to himself. The life of mere nutrition and growth may therefore be set aside. Next to this in order is what may be called the life of the senses. But even this is shared by horses, and by oxen, and by beasts. There only remains what may be described as a life of free moral action, belonging to that part of us which possesses reason, and which may possess it, either as being obedient to its commands, or as properly possessing and exercising it in consecutive thought. And, as this life can be conceived in two aspects, we will take it in its active state, for then more properly is it called life.

CHAPTER 7

EPICURUS

EPICURUS WAS THE FOUNDER of that greatly misunderstood school of philosophy known as Epicureanism. He was born in Samos, 341 B.C., and died in 270 B.C. He founded his school in Athens in 307 B.C. During the ancient times his school was looked upon as an ideal of companionship and constructive conversation. His disciples were both men and women, coming from the best families of the times, and also of the lower walks of like. Epicureanism incorporated many of the tenets of the philosophy of Democritus, that is, that the universe is purely mechanical, that everything is because of the accidental combination of the atoms in their passage through space. Epicurus did not advocate sensual pleasure, or lasciviousness; it was some of his later followers who intentionally misinterpreted his philosophy so as to dignify their own incontinence. It is to correct this misconception and at the same time to bring to your attention some of the worthy tenets of his philosophy that we give you excerpts from his writings below. Although we may not agree with them in their entirety, we cannot fail but to appreciate the fact that a great injustice was done to his philosophy by later schools designating themselves as Epicureans.

LETTER TO MENAECEUS

May no one delay to study philosophy while he is young, and when he is old, let him not become weary of the study; for no man can ever find the time unsuitable or too late to study the health of his soul. And he who asserts either that it is not yet time to philosophize, or that the hour is passed, is like a man who should say that the time is not yet come to be happy, or that it is too late. So that both young and old should study philosophy, the one in order that, when he is old, he may be young in good things through the pleasing recollection of the past, and the other in order that he may be at the same time both young and old, in consequence of his absence of fear for the future.

It is right then for a man to consider the things which produce happiness, since, if happiness is present, we have everything, and when it is absent, we do everything with a view to possess it. Now, what I have constantly recommended to you, these things I would have you do and practice, considering them to be the elements of living well. First of all, believe that God is a being incorruptible and happy; as the common opinion of the world about God dictates; and attach to your idea of him nothing which is inconsistent with incorruptibility or with happiness; and think that he is invested with everything which is able to preserve to him this happiness, in conjunction with incorruptibility. For there are gods; for our knowledge of them is indistinct. But they are not of the character which people in general attribute to them; for they do not pay a respect to them which accords with the ideas that they entertain of them. And that man is not impious who discards the god believed in by many, but he who applies to the gods the opinions entertained of them by the many. For the assertions of the many about the gods are not anticipations, but false opinions. And in consequence of these, the greatest evils which befall wicked men, and the benefits which are conferred on the good, are all attributed to the gods; for they connect all their ideas of them with a comparison of human virtues, and everything which is different from human qualities, they regard as incompatible with the divine nature.

Accustom yourself also to think death a matter with which we are not at all concerned, since all good and evil is in sensation, and since death is only the privation of sensation. On which account, the correct knowledge of the fact that death is no concern of ours, makes the mortality of life pleasant to us, insomuch as it sets forth no illimitable time, but relieves us of the longing for immortality. For there is nothing terrible in living to a man who rightly comprehends that there is nothing terrible in ceasing to live; so that he was a silly man who said that he feared death, not because it would grieve him when it was present, but because it did grieve him while it was future. For it is very absurd that that which does not distress a man when it is present, should afflict him when only expected. Therefore, the most formidable of all evils, death, is nothing to us, since, when we exist, death is not present to us; and when death is present, then we have no existence. It is no concern then either of the living or of the dead; since to the one it has no existence, and the other class has no existence itself. But people in general, at times, flee from death as the greatest of evils, and at times wish for it as a rest from the evils in life. Nor is the not living a thing feared, since living is not connected with it; nor does the wise man think not living an evil; but, just as he chooses food, not preferring that which is most abundant, but that which is nicest; so too, he enjoys time, not measuring it as to whether it is of the greatest length, but as to whether it is most agreeable. And he who enjoins a young man to live well, and an old one to die well, is a simpleton, not only because of the constantly delightful nature of life, but also because the care to live well is identical with the care to die well. And he was still wrong who said:

"Tis well to taste of life, and then when born to pass with quickness to the shades below."

For if this really was his opinion, why did he not quit life? for it was easily in his power to do so, if it really was his belief. But if he was joking, then he was talking foolishly in a case where it ought not to be allowed; and, we must recollect, that the future is not our own, nor, on the other hand, is it wholly not our own, I mean so that we can never altogether await it with a feeling of certainty that it will be, nor altogether despair of it as what will never be. And we must consider that some of the passions are natural, and some empty; and of the natural ones, some are necessary, and some merely natural. And of the necessary ones some are necessary to happiness, and others, with regard to the exemption of the body, from trouble; and others with respect to living itself; for a correct theory, with regard to these things, can refer all choice and avoidance to the health of the body and the freedom from disquietude of the soul. Since this is the end of living happily; for it is for the sake of this that we do everything, wishing to avoid grief and fear; and when once this is the case, with respect to us, then the storm of the soul is, as I may say, put an end to; since the animal is unable to go as if to something deficient, and to seek something different from that by which the good of the soul and body will be perfected.

For then we have need of pleasure when we grieve, because pleasure is not present; but when we do not grieve, then we have no need of pleasure; and on this account, we affirm, that pleasure is the beginning and end of living happily; for we have recognized this as the first good, being connate with us; and with reference to it, it is that we begin every choice and avoidance; and to this we come as if we judged of all good by passion as the standard; and, since this is the first good and connate with us, on this account we do not choose every pleasure ensuing from them; and we think many pains better than pleasures, when a greater pleasure follows them, if we endure the pain.

Every pleasure is therefore, a good on account of its own nature, but it does not follow that every pleasure is worthy of being chosen; just as every pain is an evil, and yet every pain must not be avoided.

But it is right to estimate all these things by the measurement and view of what is suitable and unsuitable; for at times we may feel the good as an evil, and at times, on the contrary, we may feel the evil as good. And, we think, contentment a great good, not in order that we may never have but a little, but in order that, if we have not much, we may make use of a little, being genuinely persuaded that those men enjoy luxury most completely who are the best able to do without it; and that every thing which is natural is easily provided, and what is useless is not easily procured. And simple flavours give as much pleasure as costly fare, when everything that can give pain, and everything feeling of want, is removed; and corn and water give the most extreme pleasure when anyone in need eats them. To accustom one's self, therefore, to simple and inexpensive habits is a great ingredient in the perfecting of health, and makes a man free from hesitation with respect to the necessary uses of life. And when we, on certain occasions, fall in with more sumptuous fare, it makes us in a better dispositon towards it, and renders us fearless with respect to fortune. When, therefore, we say that pleasure is a chief good, we are not speaking of the pleasures of the debauched man, or those which lie in sensual enjoyment, as some think who are ignorant, and who do not entertain our opinions, or else interpret them perversely; but we mean the freedom of the body from pain, and of the soul from confusion. For it is not continued drinkings and revels, or the enjoyment of female society, or feasts of fish and other such things as a costly table supplies, that make life pleasant, but sober contemplation, which examines into the reasons for all choice and avoidance, and which puts to flight the vain opinions from which the greater part of the confusion arises which troubles the soul.

CHAPTER 8

CICERO

MARCUS TULLIUS CICERO WAS born at Arpinum, Italy, in 106 B.C. He was the son of a Roman knight who possessed an estate at Arpinum and property in Rome, and most of his boyhood was spent at these two places. His studies began early; the poet Achias inspired him with a love of literature, and through the teachings of Phaedrus, he became impressed by the Epicurean philosophy. After studying dialectic, rhetoric, and law, and expressing his literary interests through verse writing and translation from Greek authors, he began his forensic career at the age of twenty-five.

The history of his life includes two unhappy marriages and features an ambitious political career during which he attained the consulship and was once sent to govern Cilicia; but his successes were interspersed with periods when he found himself deserted by his supporters, and even forced into exile. It is said that many of his failures were due to his inconsistencies, others to his excessive optimism and belief that he was exerting a good influence (as with Caesar and, later, with Octavian), which traits made him easy to deceive. Running parallel to this was his forensic career, marked by many celebrated speeches of defense and prosecution. He affected a rhythmical prose, utilized the weapon of exaggeration which was permitted to Roman orators, was a master of pathos and invective, and was considered the wit of the period. However, many thought his speeches were studded with an excessive number of jokes, and it is said that he never could resist a pun.

During his travels he studied philosophy at Athens and in Asia, and he always turned to this subject and to his writing at times of political disappointment. Many of his works show little originality, and some of them are confessedly translations and compilations; so that critics consider his chief merits the invention of a philosophical terminology for the Romans, the production of manuals which have had an enduring influence because of their beauty of style, and the famous *Letters*. These compilations are considered the chief source of information concerning that era and include 100 letters written by other persons, presenting a great variety of style.

In the year 43 B.C. his political sympathies caused him to be placed on the proscribed list and on December 7 he was beheaded by Mark Anthony's soldiers. He had attempted to escape on a ship but, being cast back by unfavorable winds, he returned to his villa saying: "Let me die in the country which I have often saved." His hands and head were nailed to the rostra in Rome after Antony's wife had thrust a hairpin through the tongue!

Cicero's literary works are classed as rhetorical, oratorical, philosophic, political, and epistolary. Aside from his *Letters*, some of his best liked works are: *De Republica* (which contains the famous, prophetic *Dream of Scipio* often compared to the *Vision of Er* in Plato's *Republic*), *De Finibus, De Natura Deorum*, and *Tusculanae Disputationes*. From this last-named work we have chosen a few quotations on the subject of immortality from the section entitled "The Contempt of Death."

THE CONTEMPT OF DEATH

The first thing, then, is to inquire what death, which seems to be so well understood, really is; for some imagine death to be the departure of the soul from the body; others think that there is no such departure, but that soul and body perish together, and that the soul is extinguished with the body. Of those who think that the soul does depart from the body, some believe in its immediate dissolution; others fancy that it continues to exist for a time; and others believe that it lasts forever. . . .

But the greatest proof of all is, that nature herself gives a silent judgment in favor of the immortality of the soul, inasmuch as all are anxious, and that to a great degree, about the things which concern futurity;—"One plants that future ages shall enjoy," as Statius said in his Synephebi. What is his object in doing so, except that he is interested in posterity? Shall the industrious husbandman, then, plant trees the fruit of which he shall never see? and shall not the great man found laws, institutions, and a republic? What does the procreation of children imply—and our care to continue our names—and our adoptions—and our scrupulous exactness in drawing up wills—and the inscriptions on monuments, and panegyrics, but that our thoughts run on futurity? There is no doubt but a judgment may be formed of nature in general, from looking at each nature in its most perfect specimen; and what is a more perfect specimen of man, than those who look on themselves as born for the assistance, the protection, and the preservation of others? . . .

What will you say? what do you imagine that so many and such great men of our republic, who have sacrificed their lives for its good, expected? Do you believe that they thought that their names should not continue beyond their lives? None ever encountered death for their country, but under a firm persuasion of immortality! Themistocles might have lived at his ease; so might Epaminondas; and, not to look abroad and amongst the ancients for instances, so might I myself. But somehow or other, there clings to our minds a certain presage of future ages; and this both exists most firmly and appears most clearly, in men of the loftiest genius and greatest souls. Take away this, and who would be so mad as to spend his life among toils and dangers? I speak of those in power. What are the poet's views but to be ennobled after death? . . . Why do I mention poets? the very mechanics are desirous of fame after death. Why did Phidias include a likeness of himself in the shield of Minerva, when he was not allowed to inscribe his name on it? What do our philosophers think on the subject? Do they not put their names to those very books which they write on the contempt of glory? If, then, universal consent is the voice of nature, and if it is the general opinion everywhere, that those who have quitted this life are still interested in something; we must also subscribe to that opinion. And if we think that men of the greatest abilities and virtue see most clearly into the power of nature, because they themselves are her most perfect work; it is very probable that, as every great man is especially anxious to benefit posterity, there is something of which he himself will be sensible after death.

But as we are led by nature to think there are gods, and as we discover, by reason, of what description they are, so by the consent of all nations, we are induced to believe that our souls survive; but where their habitation is, and of what character they eventually are, must be learned from reason. . . .

But I return to the ancients. They scarcely ever gave any reason for their opinion but what could be explained by numbers or definitions. It is reported of Plato, that he came into Italy to make himself acquainted with the Pythagoreans; and that when there, amongst others, he made an acquaintance with Archytas and Timaeus, and learned from them all the tenets of the Pythagoreans; and that he not only was of the same opinion with Pythagoras concerning the immortality of the soul, but that he also brought reasons in support of it . . . .

What kind of sight do you imagine that will be, when the whole earth is laid open to our view? And that, too, not only in its position, form, and boundaries, nor those parts of it only which are habitable, but those also that lie uncultivated, through the extremities of heat and cold to which they are exposed; for not even now is it with our eyes that we view what we see, for the body itself has no senses; but (as the naturalists, aye, and even the physicians assure us, who have opened our bodies, and examined them), there are certain perforated channels from the seat of the soul to the eyes, ears, and nose; so that frequently, when either prevented by meditation, or the force of some bodily disorder, we neither hear nor see, though our eyes and ears are open, and in good condition; so that we may easily apprehend that it is the soul itself which sees and hears, and not those parts which are, as it were, but windows to the soul; by means of which, however, she can perceive nothing, unless she is on the spot, and exerts herself. How shall we account for the fact, that by the same power of thinking we comprehend the most different things; as color, taste, heat, smell, and sound? which the soul could never know by her five messengers, unless everything was referred to her, and she were the sole judge of all. And we shall certainly discover these things in a more clear and perfect degree when the soul is disengaged from the body, and has arrived at that goal to which nature leads her, for at present, notwithstanding nature has contrived, with the greatest skill, those channels which lead from the body to the soul, yet they are, in some way or other, stopped up with earthy and concrete bodies; but when we shall be nothing but soul, then nothing will interfere to prevent our seeing everything in its real substance, and in its true character. . . .

Not that I see any reason why the opinion of Pythagoras and Plato may not be true; but even although Plato were to have assigned no reason for his opinion (observe how I esteem the man), the weight of his authority would have borne me down; but he has brought so many reasons, that he appears to me to have endeavored to convince others, and certainly to have convinced himself.

But there are many who labor on the other side of the question, and condemn souls to death, as if they were criminals capitally convicted; nor have they any other reason to allege why the immortality of the soul appears to them to be incredible, except that they are not able to conceive what sort of thing the soul can be when disentangled from the body; just as if they could really form a correct idea as to what sort of thing it is, even when it is in the body; what its form, and size, and abode are; so that were they able to have a full view of all that is now hidden from them in a living body, they have no idea whether the soul would be discernible by them, or whether it is of so fine a texture that it would escape their sight. Let those consider this, who say they are unable to form any idea of the soul without the body, and then they will see whether they can form any adequate idea of what it is when it is in the body. For my own part, when I reflect on the nature of the soul, it appears to me a far more perplexing and obscure question to determine what is its character while it is in the body, a place which, as it were, does not belong to it, than to imagine what it is when it leaves it, and has arrived at the free aether, which is, if I may say so, its proper, its own habitation. . . .

I think I can account for the manner in which the blood, and bile, and phlegm, and bones, and nerves, and veins, and all the limbs, and the shape of the whole body, were put together and made; aye, and even as to the soul itself, were there nothing more in it than a principle of life, then the life of a man might be put upon the same footing as that of a vine or any other tree, and accounted for as caused by nature; for these things, as we say, live. Besides, if desires and aversions were all that belonged to the soul, it would have them only in common with the beasts; but it has, in the first place, memory, and that, too, so infinite, as to collect an absolute countless number of circumstances, which Plato will have to be a recollection of a former life; for in that book which is inscribed Menon, Socrates asks a child some questions in geometry, with reference to measuring a square; his answers are such as a child would make, and yet the questions are so easy, that while answering them, one by one, he comes to the same point as if he had learned geometry. From whence Socrates would infer, that learning is nothing more than recollection; and this topic he explains more accurately, in the discourse which he held the very day he died; for he there asserts that any one who seeming to be entirely illiterate, is yet able to answer a question well that is proposed to him, does in so doing manifestly show that he is not learning it then, but recollecting it by his memory. Nor is it to be accounted for in any other way, how children come to have notions of so many and such important things, as are implanted, and as it were sealed up in their minds, (which the Greeks call *ennoiai*) unless the soul before it entered the body had been well stored with knowledge.

CHAPTER 9

LUCRETIUS

TITUS LUCRETIUS CARUS WAS one of the most forceful writers in verse ever known. He was born of a noble family approximately 98 B.C., and died about 55 B.C. Very little is known of his life except that he was an outstanding pupil of the eminent philosopher, Epicurus. His philosophy is not merely a recital of the Epicurean philosophy, but consists of his own ideas as well. He, like his master, accepted the Democritean atomic philosophy which stated that all matter consisted of small seeds or particles which united by compact, and the quantity of them determined the nature of matter. The universe, then, was accounted for according to a mechanical principle—the mere coming together or combining of these little particles. There was no intelligent cause for all. The particles in themselves were purposeless. In a very splendid treatise Lucretius applies this theory to life and to man. The title of his thesis is *On Immortality*. We quote a portion of it below. It will prove very profitable to you. It is immaterial whether you agree with his conception or not; you will benefit by the logical presentation of his arguments and admire the forceful arrangement of his thoughts.

ON IMMORTALITY

And now attend. That thou mayest understand that living creatures have minds, and subtle souls born and perishable, I will proceed to arrange verses worthy of thy life *and virtues, verses* collected during a long time, and prepared with sweet labor. *And* thou, *my friend*, take care to include both of them under one name, *whichsoever of the two I may use*; and, for example when I proceed to speak of the soul, teaching that it is mortal, suppose that I also speak of the mind; inasmuch as they are one by mutual *combination*, and their substance is united.

In the first place, since I have shown that the *soul, being* subtle, consists of minute particles, and is composed of much smaller atoms than the clear fluid of water, or mist, or smoke; (for it far surpasses *those bodies* in susceptibility-of-motion, and is more readily impelled when acted upon from a slight cause; inasmuch as *both the mind and soul* are moved by the *mere* images of smoke and mist; as when, lulled in sleep, we see high altars exhale with vapor, and carry up smoke; since doubtless these phantasms are produced in us;) now, therefore, *I say*, since, when vessels are broken to pieces, you see water flow about, and *any other* liquid runs away; and since, *also*, mist and smoke disperse into the air; *you must* conclude that the soul is likewise scattered abroad, and is dissipated much sooner *than mist and smoke*, and more easily resolved into *its* original elements, when it *has* once *been* withdrawn from the body of a man, *and* has taken its departure. For how can you believe that this *soul* can be held together by any *combination* of air, when the body itself (which is, as it were, its vessel) cannot contain it, *if it be* convulsed by any violence, or rendered thin *and weak* by blood being taken from the veins? How can *that air which is* more rare than our body confine it?

Besides, we observe that the mind is produced together with the body, and grows up along *with it*, and waxes old at the same time *with it*. For as children wander *and totter* about with a weak and tender body, so the subtle sense of the mind follows *and corresponds to the weakness of their frame*. Then, when their age has grown up in robust vigor, their understanding is also greater, and their strength of mind more enlarged. Afterwards, when the body is shaken by the prevailing power of time, and, the strength being depressed, the limbs have sunk *into* infirmity, the undersanding *then* halts, the tongue and the mind lose their sense, all *parts* fail and fade away at once. It is therefore natural that the whole substance of the soul should be dissolved, as smoke, into the sublime air of heaven; since we see that it is produced together with the body, and grows up together *with it*, and both, as I have shown, overcome by age, decay in concert.

To this is added, that as we observe the body itself to-be-subject-to violent diseases and severe pain, so *we see* the mind *to be susceptible of* sharp cares, and grief, and fear, For which cause it is reasonable that it should also be a partaker of death.

Moreover the mind, in diseases of the body, often wanders distracted; for it loses its faculties, and utters senseless words; and sometimes, by a heavy lethargy, is born down into a deep and eternal sleep, the eyes and the nodding-head sinking; hence it neither hears the voice, nor can distinguish the countenances, of those who stand around recalling it to life, bedewing their faces and cheeks with tears. Wherefore you must necessarily admit that the mind is also dissolved, since the contagion of disease penetrates into it. For pain and disease are each the fabricator of death; *a truth* which we have been taught by the destruction of many *millions* in past times.

Further, when the violent power of wine has penetrated the heart of men, and its heat, being distributed, has spread into the veins, a heaviness of the limbs follows, the legs of the tottering person are impeded, the tongue grows torpid, the mind is, *as it were*, drowned; noise, hiccups, and quarrels arise, and other things of this kind, whatever are consequent *on intoxication*. Why do these *effects* happen, unless because the vehement force of the wine has exerted-its-customary-power to disturb the soul *as it is diffused* through the body itself? But whatsoever things can be *thus* disturbed and obstructed *in their operations*, show, that if a cause somewhat stronger shall spread within *them, the consequence* will be that they must perish, deprived of *all* future existence.

Further, the animated powers of the body and mind are vigorous, and enjoy life, only when joined with one another; for neither can the nature *or substance* of the mind, without the body, alone, *and* of itself, produce vital motions; nor again *can* the body, deprived of the soul, continue *its state of existence*, and use its faculties. Just, for example, as the eye itself, torn from its roots, can discern no object apart from the whole body, so the mind or soul seems to have no power in itself; evidently because when mingled throughout the veins and viscera, throughout the nerves and bones, they are held-in-close-confinement by the whole body, and their primary-particles, not being free, cannot fly asunder to great distances; consequently, being thus confined, they move with sensitive motions, with which, after death, when cast forth beyond the body into the air of heaven, they cannot move; for this very reason, that they are not held-confined in a similar manner. For *surely* the air forms body and soul, if the soul shall be able to keep itself together *in the air*, and to contain itself for *exerting* those motions, which it before exercised amidst the nerves, and in the body itself. On which account, *I say* again and again, you must necessarily admit that when the whole enclosure of the body is dissolved, and the vital breath cast forth, the sentient-existence of the mind and the soul is dissolved; since there is common cause *and like fate* to both.

Besides, when the body cannot bear the dissociation of the soul, without putrifying with offensive odor, why do you doubt that the essence of the soul, rising from the depths and innermost part *of the body*, has passed forth, *and has been* diffused abroad like smoke? and that for this reason the body, decaying with so great a dissolution, has utterly fallen away, because the foundations have been removed from their place, and the spirits pass out through the limbs, and through all the windings of the passages and ducts that are in the body? So that you may understand from many considerations that the nature *or substance* of the soul, being departed, has gone out through the members *of the body*, and that it was dissevered within the body itself, before gliding outwards, it flowed forth into the air of heaven.

Moreover, whilst the soul dwells within the bounds of life, it yet frequently, when it has received a shock from some cause, seems to pass away, and *presents the appearance* that the mind is let loose from the whole body; and the countenance *then* seems to become inanimate as at the last hour, and all the relaxed members to fail the languid frame. Such is the case, when it is said that the mind has been damaged, or the vital power has suffered syncope; while *all* is trepidation, and all are anxious to recover the last link of life. For then all the mind, and power of the soul, are shaken; and these, *it is evident,* sink with the body itself; *so* that a cause of somewhat greater force may bring *them* to dissolution.

Why *then* do you doubt, but that, *at the hour of death, the soul* driven forth at length, weak *and helpless*, out of the body, and being in the open air, with its covering removed, can not only not endure throughout all time, but cannot even maintain-its-existence for the smallest space whatsoever?

CHAPTER 10

SENECA, THE YOUNGER

LUCIUS ANNAEUS SENECA, A Roman Stoic philosopher, was born at Corduba, Spain, about 4 B.C. He studied rhetoric and philosophy, and under the tutelage of the Pythagorean, Sotion, he acquired a great admiration for Pythagoras and his teachings. Later he left Sotion to follow the teachings of Attalus the Stoic. He attained recognition at the bar but gave up these activities because of fear of Caligula's jealousy. In 41 A.D. Claudius ordered his exile to Corsica at the instigation of Messalina. After eight years he was recalled by Agrippina, wife of Claudius, and was given the position of tutor to Nero, her son. Seneca had considerable influence on the headstrong Nero during his youthful years and due to the young emperor's extravagant bounty, Seneca accumulated a vast fortune. Seneca's good fortune caused many enemies, who made it a point to impress upon Nero his increasing power and popularity and to arouse in the emperor as much jealousy toward Seneca as possible. Seneca sensed the pending difficulties and offered to return the Emperor's gifts and retire on a small allowance. Nero refused, and Seneca withdrew into private life and was seldom seen in Rome. Nero failed in an attempt to have him poisoned, but shortly afterwards Seneca was implicated in the Piso conspiracy and sentenced to end his own life. This he did by bleeding himself to death in 65 A.D.

The most important writings of Seneca consist of his philosophical, moral, and ethical discourses. In his day he enjoyed an unrivaled popularity as a writer. Besides his epistles and short treatises, he wrote seven books and ten tragedies, the latter having a great influence on Renaissance, French, and English classical drama.

ON LEISURE

What I have to say (on Leisure) I shall develop under two heads, showing, first, that it is possible for a man to surrender himself wholly to the contemplation of truth, to search out the art of living, and to practice it in retirement, even from his earliest years; secondly, that, when a man has now earned release from public service and his life is almost over, it is possible that he may with perfect justice do the same thing and turn his mind to quite different activities, after the manner of the Vestal virgins, whose years are allotted to varied duties while they are learning to perform the sacred rites, and when they have learned, they begin to teach.

I shall show, too, that the Stoics also accept this doctrine, not because I have made it my rule to set up nothing contrary to the teaching of Zeno or Chrysippus, but because the matter itself suffers me to adopt their opinion; for if a man always follows the opinion of one person, his place is not in the senate, but in a faction. Would that all things were now understood, that truth were uncovered and revealed, and that we never altered our mandates! As it is, we are in search of truth in company with the very men that teach it.

The two sects, the Epicureans and the Stoics, are at variance as in most things, in this matter also; they both direct us to leisure, but by different roads. Epicurus says: "The wise man will not engage in public affairs except in an emergency." Zeno says: "He will engage in public affairs unless something prevents him." The one seeks leisure by fixed purpose, the other for a special cause; but the term "cause" has here broad application. If the state is too corrupt to be helped, if it is wholly dominated by evils, the wise man will not struggle to purpose, nor spend himself when nothing is gained. If he is lacking in influence or power and the state is unwilling to accept his services, if he is hampered by ill health, he will not enter upon a course for which he knows he is unfitted, just as he would not launch upon the sea a battered ship, just as he would not enlist for a service in the army if he were disabled. Consequently, it is also possible that a man whose fortunes are still unharmed may establish himself in a safe retreat before he experiences any of the storms of life, and thenceforth devote himself to the liberal studies and demand uninterrupted leisure to cultivate the virtues, which even those who are most retired are able to practice. It is of course required of a man that he should benefit his fellow-men—many, if he can, if not, a few; if not a few, those who are nearest; if not these, himself. For when he renders himself useful to others, he engages in public affairs. Just as the man that chooses to become worse injures not only himself but all those whom, if he had become better, he might have benefitted, so whoever wins the approval of himself benefits others by the very fact that he prepares what will prove beneficial to them.

Let us grasp the idea that there are two commonwealths—the one, a vast and truly common state, which embraces alike gods and men, in which we look neither to this corner of earth nor to that, but measure the bounds of our citizenship by the path of the sun; the other, the one to which we have been assigned by the accident of birth. This will be the commonwealth of the Athenians or of the Carthaginians, or of any other city that belongs, not to all, but to some particular race of men. Some yield service to both commonwealths at the same time—to the greater and to the lesser—some only to the lesser, some only to the greater. This greater commonwealth we are able to serve even in leisure—nay, I am inclined to think, even better in leisure. . . .

And with what thought does the wise man retire into leisure? In the knowledge that there also he will be doing something that will benefit posterity. Our school at any rate is ready to say that both Zeno and Chrysippus accomplished greater things than if they had led armies, held public office, and framed laws. The laws they framed were not for one state only, but for the whole human race. Why, therefore, should such leisure as this not be fitting for the good man, who by means of it may govern the ages to come, and speak, not to the ears of the few, but to the ears of all men of all nations, both those who now are and those who shall be? In brief, I ask you whether Cleanthes and Chryssipus and Zeno lived in accordance with their teachings. Undoubtedly you will reply that they lived just as they taught that men ought to live. And yet no one of them governed a state. You reply: "They had neither the fortune nor the rank which ordinarily admit one to the management of public affairs." But, nevertheless, they did not lead a life of sloth; they found a way to make their own repose a greater help to mankind than all the pother and sweat of others. Therefore, though they played no public part, they none the less have been thought to have played a great part.

Moreover, there are three kinds of life, and it is a common question as to which of them is best. One is devoted to pleasure, a second to contemplation, a third to action. Having first put away our strife and having put away the hatred which we have relentlessly declared against those who pursue ends different from ours, let us see how all these, under different names, come to the same thing. For he who sanctions pleasure is not without contemplation, nor he who surrenders to contemplation without pleasure, nor is he whose life is devoted to action without contemplation. But you say: "Whether something is a chief aim or is merely attached to some other chief aim makes a very great difference." Yes, grant that there is a huge difference, nevertheless the one does not exist without the other. That man is not given to contemplation without action, nor this one to action without contemplation, nor does that third one—concerning whom we have agreed to form a bad opinion—give sanction to idle pleasure, but to the pleasure that he renders stable for himself by his reason; thus even this pleasure-loving sect is itself committed to action. Clearly is it committed to action! Since Epicurus himself declares that he will at times withdraw from pleasure, will even seek pain if he forsees that he will either repent of pleasure, or will be able to substitute a lesser pain for one that is greater. And what is my purpose in stating these things? To make it clear that contemplation is favoured by all. Some men make it their aim; for us it is a roadstead, but not the harbour.

Add, further, that on the authority of Chrysippus a man has a right to live a life of leisure; I do not mean, that he may tolerate leisure, but that he may choose it. Our school refuses to allow the wise man to attach himself to any sort of state. But what difference does it make in what manner the wise man arrives at leisure—whether because no state is available to him or because he is not available to the state—if he is nowhere to find a state? Besides, no state will ever be available to the fastidious searcher. I ask you to what state should the wise man attach himself? To that of the Athenians, in which Socrates was sentenced to death, from which Aristotle fled to avoid being sentenced? in which all the virtues are crushed by envy? Surely you will say that no wise man will wish to attach himself to this state. Shall the wise man, then, attach himself to the state of the Carthaginians, in which faction is always rife and all the best men find "freedom" their foe, in which justice and goodness have supreme contempt, and enemies are treated with inhuman cruelty and fellow citizens like enemies? From this state also will he flee. If I should attempt to enumerate them one by one, I should not find a single one which could tolerate the wise man or which the wise man could tolerate. But if that state which we dream of can nowhere be found, leisure begins to be a necessity for all of us, because the one thing that might have been preferred to leisure nowhere exists. If anyone says that the best life of all is to sail the sea, and then adds that I must not sail upon a sea where shipwrecks are a common occurrence and there are often storms that sweep the helmsman in an adverse direction, I conclude that this man, although he lauds navigation, really forbids me to launch my ship.

CHAPTER 11

EPICTETUS

EPICTETUS, ONE OF THE great authorities on Stoic philosophy, was born at Hieropolis, Phrygia, about the middle of the first century A.D. In his youth he was a slave in Rome, and it is said that his lameness—the only physical characteristic of his which was recorded—was caused by the cruelty of his master. Origen records that, when he was being tortured, Epictetus said calmly: "You will break my leg," and when it broke, just as quietly, "Didn't I tell you?"

He learned the principles of Stoic philosophy through the lectures of the famous Roman teacher Musonius Rufus. After his emancipation he taught this system in Rome until about the year 90 A.D. when the emperor Domitian banished all philosophers. Epictetus then established a school at Nicopolis, in Epirus, and gained a high reputation through his discourses. He taught only by means of conversation and lectures, and of these we have only the portions which were written down by his favorite pupil, Arrian, who is also known as the biographer of Alexander the Great.

The time and manner of Epictetus' death are unknown but one feels that they were unimportant even to himself, and that, with the same attitude which he had fostered in his pupils, he departed assured: ". . . that He is sounding the retreat, that He hath opened the door, and is saying to thee, Come!"

GOLDEN SAYINGS

What saith God?—"Had it been possible, Epictetus, I would have made both that body of thine and thy possessions free and unimpeded, but as it is, be not deceived:—it is not thine own; it is but finely tempered clay. Since then this I could not do, I have given thee a portion of Myself, in the power of desiring and declining and of pursuing and avoiding, and in a word the power of dealing with the things of sense. And if thou neglect not this, but place all that thou hast therein, thou shalt never be let or hindered; thou shalt never lament; thou shalt not blame or flatter any. What then? Seemeth this to thee a little thing?"—God forbid!—"Be content then therewith!"

And so I pray the Gods.

IX. If a man could be thoroughly penetrated, as he ought, with this thought, that we are all in an especial manner sprung from God, and that God is the Father of men as well as of Gods, full surely he could never conceive aught ignoble or base of himself. Whereas if Caesar were to adopt you, your haughty looks would be intolerable; will you not be elated at knowing that you are the son of God? Now however it is not so with us: but seeing that in our birth these two things are commingled—the body which we share with the animals, and the Reason and Thought which we share with the Gods, many decline towards this unhappy kinship with the dead, few rise to the blessed kinship with the Divine. Since then everyone must deal with each thing according to the view which he forms about it, those few who hold that they are born for fidelity, modesty, and unerring sureness in dealing with the things of sense, never conceive aught base or ignoble of themselves: but the multitude the contrary. Why, what am I?—A wretched human creature; with this miserable flesh of mine. Miserable indeed! but you have something better than that paltry flesh of yours. Why then cling to the one, and neglect the other?

XIII. But God hath introduced Man to be a spectator of Himself and of His works; and not a spectator only, but also an interpreter of them. Wherefore it is a shame for man to begin and to leave off where the brutes do. Rather he should begin there, and leave off where Nature leaves off in us: and that is at contemplation, and understanding, and a manner of life that is in harmony with herself.

See then that ye die not without being spectators of these things.

XVI. He that hath grasped the administration of the World, who hath learned that this Community, which consists of God and men, is the foremost and mightiest and most comprehensive of all:—that from God have descended the germs of life, not to my father only and father's father, but to all things that are born and grow upon the earth, and in an especial manner to those endowed with Reason (for those only are by their nature fitted to hold communion with God, being by means of Reason conjoined with Him)—why should not such an one call himself a citizen of the world? Why not a son of God? Why should he fear aught that comes to pass among men? Shall kinship with Caesar, or any other of the great at Rome, be enough to hedge men around with safety and consideration, without a thought of apprehension: while to have God for our Maker, and Father, and Kinsman, shall not this set us free from sorrows and fears?

XX. Seeing this then and noting well the faculties which you have, you should say,—"Send now, O God, any trial that Thou wilt; lo, I have means and powers given me by Thee to acquit myself with honour through whatever comes to pass!"—No; but there you sit, trembling for fear certain things should come to pass, and moaning and groaning and lamenting over what does come to pass. And then you upbraid the Gods. Such meanness of spirit can have but one result—impiety.

Yet God has not only given us these faculties by means of which we may bear everything that comes to pass without being crushed or depressed thereby; but like a good King and Father, He has given us this without let or hindrance, placed wholly at our own disposition, without reserving to Himself any power of impediment or restraint. Though possessing all these things free and all your own, you do not use them! you do not perceive what it is you have received nor whence it comes, but sit moaning and groaning; some of you blind to the Giver, making no acknowledgement to your Benefactor; others basely giving themselves to complaints and accusations against God.

Yet what faculties and powers you possess for attaining courage and greatness of heart, I can easily show you; what you have for upbraiding and accusation, it is for you to show me!

XXX. You must know that it is no easy thing for a principle to become a man's own, unless each day he maintain it and hear it maintained, as well as work it out in life.

XXXV. When we are invited to a banquet we take what is set before us; and were one to call upon his host to set fish upon the table or sweet things, he would be deemed absurd. Yet in a word, we ask the Gods for what they do not give; and that, although they have given us so many things!

XLI. What you shun enduring yourself, attempt not to impose on others. You shun slavery—beware of enslaving others! If you can endure to do that, one would think you had been once upon a time a slave yourself. For Vice has nothing in common with virtue, nor Freedom with slavery.

LXV. When a youth was giving himself airs in the Theatre and saying, "I am wise, for I have conversed with many wise men." Epictetus replied, "I too have conversed with many rich men, yet I am not rich!"

LXXXVII. The husbandman deals with land; physicians and trainers with the body; the wise man with his own Mind.

XCVIII. Nevertheless a man should also be prepared to be sufficient unto himself—to dwell with himself alone, even as God dwells with Himself alone, shares His repose with none, and considers the nature of His own administration, intent upon such thoughts as are meet unto Himself. So should we also be able to converse with ourselves, to need none else beside, to sigh for no distraction, to bend our thoughts upon the Divine Administration, and how we stand related to all else; to observe how human accidents touched us of old, and how they touch us now; what things they are that still have power to hurt us, and how they may be cured or removed; to perfect what needs perfecting as Reason would direct.

CXXXIV. To a good man there is no evil, either in life or death. And if God supply not food, has He not, as a wise Commander, sounded the signal for retreat and nothing more? I obey, I follow—speaking good of my Commander, and praising His acts. For at His good pleasure I came; and I depart when it pleases Him; and while I was yet alive that was my work, to sing praises unto God!

CXXXV. Reflect that the chief source of all evils to Man, and of baseness and cowardice, is not death, but the *fear* of death.

Against this fear then, I pray you, harden yourself; to this let all your reasonings, your exercises, your reading tend. Then shall you know that thus alone are men set free.

CHAPTER 12

MARCUS AURELIUS

MARCUS AURELIUS ANTONINUS WAS born in Rome, 121 A.D., on the 26th of April, in the Verus family. He was adopted by the emperor as a youth. He was carefully brought up in an excellent home environment; he refers to this in many of his writings and was appreciative of the cultural background it gave him. Like the youth of the higher classes of society in Rome, he tried poetry and studied rhetoric. When he was but eleven years of age, he assumed the dress of the philosophers—something plain and coarse, to show one's disregard for the outer things of life and concern for knowledge and wisdom instead. He studied extremely hard and laboriously, to the extent that it injured his health.

He eventually abandoned both poetry and rhetoric, and in fact all studies except law and philosophy. Later in life he became a member of the sect of the Stoics, and his thinking was molded under the excellent teachers of that school.

The Emperor Antoninus Pious died in March, 161 A.D.; Marcus Aurelius Antoninus associated himself with the other adopted son of Antoninus Pious, L. Ceionius Commodus, and both became emperors together for quite a period. Later, when he was ruling alone, he traveled through Syria and Egypt, and on his return to Italy visited Athens and was initiated into the Eleusinian mysteries.

It was not until Marcus Aurelius' successors became Christian that such strong opposition was aroused against the "pagan beliefs" that the old mysteries of the Classical world were terminated. It must be realized, however, that much of the persecution that the Christians had received during the early Roman Empire was due to the fact that they, themselves, brought war against Roman beliefs and customs. This naturally agitated the Roman believers against them. While the philosopher-emperor did uphold the laws against the Christians made by his predecessors, he did not unleash any persecutions against them.

The following excerpts are taken from what has been called *The Emperor's Reflections* or *The Meditations of Marcus Aurelius.* We are indebted to George Long for his work on this subject.

MEDITATIONS

We ought to understand not only that our life is daily wasting away and a smaller part of it is left, but another thing also must be taken into the account, that if a man should live longer, it is quite uncertain whether the understanding will still continue sufficient for the comprehension of things, and retain the power of contemplation which strives to acquire the knowledge of the divine and the human. For if he shall begin to fall into dotage, perspiration and nutrition and imagination and appetite, and whatever else there is of the kind, will not fail; but the power of making use of ourselves, and filling up the measure of our duty, and clearly separating all appearances, and considering whether a man should now depart from life, and whatever else of the kind absolutely requires a disciplined reason—all this is already extinguished. We must make haste, then, not only because we are daily nearer to death, but also because the conception of things and the understanding of them cease first.

We ought to observe also that even the things which follow after the things which are produced according to nature contain something pleasing and attractive. For instance, when bread is baked some parts are split at the surface, and these parts which thus open, and have a certain fashion contrary to the purpose of the baker's art, are beautiful in a manner, and in a peculiar way excite a desire for eating. And again, figs when they are quite ripe, gape open; and in the ripe olives the very circumstances of their being near to rottenness adds a peculiar beauty to the fruit. And the ears of corn bending down, and the lion's eyebrows, and the foam which flows from the mouth of wild boars, and many other things,—though they are far from being beautiful if a man should examine them severally,—still, because they are consequent upon the things which are formed by nature, help to adorn them, and they please the mind; so that if a man should have a feeling and deeper insight with respect to the things which are produced in the universe, there is hardly one of those which follow by way of consequence which will not seem to him to be in a manner disposed so as to give pleasure. And so he will see even the real gaping jaws of wild beasts with no less pleasure than those which painters and sculptors show by imitation; and in an old woman and an old man he will be able to see a certain maturity and comeliness; and the attractive loveliness of young persons he will be able to look on with chaste eyes; and many such things will present themselves, not pleasing to every man, but to him only who has become truly familiar with Nature and her works.

Hippocrates, after curing many diseases, himself fell sick and died. The Chaldaei foretold the deaths of many, and then fate caught them too. Alexander and Pompeius and Caius Caesar, after so often completely destroying whole cities, and in battle cutting to pieces many ten thousands of cavalry and infantry, themselves too at last departed from life. Heraclitus, after so many speculations on the conflagration of the universe, was filled with water internally and died smeared all over with mud. And lice destroyed Democritus; and other lice killed Socrates. What means all this? Thou hast embarked, thou hast made the voyage, thou art come to shore; get out. If indeed to another life, there is no want of gods, not even there; but if to a state without sensation, thou wilt cease to be held by pains and pleasures, and to be a slave to the vessel, which is as much inferior as that which serves it is superior: for the one is intelligence and deity; the other is earth and corruption.

Do not waste the remainder of thy life in thoughts about others, when thou dost not refer thy thoughts to some object of common utility. For thou losest the opportunity of doing something else when thou hast such thoughts as these,—What is such a person doing, and why, and what is he saying, and what is he thinking of, and what is he contriving, and whatever else of the kind makes us wander away from the observation of our own ruling power. We ought then to check in the series of our thoughts everything that is without a purpose and useless, but most of all the overcurious feeling and the malignment; and a man should use himself to think of those things only about which if one should suddenly ask, What has thou now in thy thoughts? with perfect openness thou mightest immediately answer, This or That; so that from thy words it should be plain that everything in thee is simple and benevolent, and such as befits a social animal, and one that cares not for thoughts about pleasure or sensual enjoyments at all, nor has any rivalry or envy and suspicion, or anything else for which thou wouldst blush if thou shouldst say that thou hadst it in thy mind. For the man who is such, and no longer delays being among the number of the best, is like a priest and minister of the gods, using too the (deity) which is planted within him, which makes the man uncontaminated by pleasure, unharmed by any pain, untouched by any insult, feeling no wrong, a fighter in the noblest fight, one who cannot be overpowered by any passion, dyed deep with justice, accepting with all his soul everything which happens and is assigned to him as his portion; and not often, nor yet without great necessity and for the general interest, imagining what another says, or does, or thinks. For it is only what belongs to himself that he makes the matter for his activity; and he constantly thinks of that which is allotted to himself out of the sum total of things, and he makes his own acts fair, and he is persuaded that his own portion is good. For the lot which is assigned to each man is carried along with him and carries him along with it. And he remembers also that every rational animal is his kinsman, and that to care for all men is according to man's nature; and a man should hold on to the opinion not of all, but of those only who confessedly live according to nature. But as to those who live not so, he always bears in mind what kind of men they are both at home and from home, both by night and by day, and what they are, and with what men they live an impure life. Accordingly, he does not value at all the praise which comes from such men, since they are not even satisfied with themselves.

Labor not unwillingly, nor without regard to the common interest, nor without due consideration, nor with distraction; nor let studied ornament set off thy thoughts, and be not either of many words or busy about too many things. And further, let the deity which is in thee be the guardian of a living being, manly and of ripe age, and engaged in matter political, and a Roman, and a ruler, who has taken his post like a man waiting for the signal which summons him from life, and ready to go, having need neither of oath nor of any man's testimony. Be cheerful also, and seek not external help nor the tranquillity which others give. A man must stand erect, not be kept erect by others.

CHAPTER 13

PLOTINUS

PLOTINUS WAS BORN IN Lycopolis, Egypt, 205 A.D., supposedly of a Roman family, but in keeping with his indifference to things human, "being ashamed almost to live in a body," he never divulged his parentage. He believed the body to be only a faint image of existence, therefore he never allowed his birthday to be celebrated, nor would he allow a painter or sculptor to perpetuate his features. To him his body was utterly contemptible and therefore not worthy of a physician's care when ill; he ate sparingly and eliminated meat from his diet. He was twenty-eight years old before philosophy seemed to interest him, and while in Alexandria he found a congenial teacher in Ammonius Saccas. After ten years of ardent study and work, he become the chief representative and author of that school with the reputation of being an independent thinker who developed his theories to the fullest extent. In 242 A.D. he accompanied Emperor Gordianus' expedition to Persia, where he hoped to devote himself to the philosophy of the East. The emperor was assassinated in Mesopotamia, however, and with some difficulty Plotinus escaped to Antioch. In 244 A.D. he went to Rome, where he spent the remainder of his life. There he opened a school which was attended by men and women of the highest circles. His instruction was mostly oral, for he seemed little interested in conserving his teachings; however, to his pupil, Porphyry, we owe the preservation of his scattered lectures and teachings. Plotinus died after a long illness at the age of sixty-six, in 270 A.D. He was a man of saintly character and very attractive personality. He made no enemies and was reverenced by all who knew him. His philosophy tried to combine the systems of Anaxagoras, Parmenides, the Pythagoreans, Plato, Socrates, and the Stoics into one, and its influence upon modern philosophy is remarkable.

THE ONE IN ITSELF

What then will *the one* be; and what nature will it possess? Or may we not say that it is not at all wonderful, it should not be easy to tell what it is, since neither is it easy to tell what being is, or what form is. But our knowledge is fixed in forms. When, however, the soul directs its attention to that which is formless, then being unable to comprehend that which is not bounded, and as it were impressed with forms by a former of a various nature, it falls from the apprehension of it, and is afraid it will possess (nothing from the view). Hence, it becomes weary in endeavours of this kind, and gladly descends from the survey frequently falling from all things, till it arrives at something sensible, and as it were rests in a solid substance; just as the sight also, when wearied with the perception of small objects, eagerly converts itself to such as are large. When, however, the soul wishes to perceive by itself, and sees itself alone, then in consequence of being one with the object of its perception, it does not think that it yet possesses that which it investigates, because it is not different from that which it investigates, because it is not different from that which it intellectually perceives. At the same time, it is requisite that he should act in this manner, who intends to philosophize about *the one*. Since, therefore, that which we investigate is one, and we direct our attention to the principle of all things, to *the good*, and the first, we ought not to be far removed from the natures which are about the first of things, nor fall from them to the last of all things, but proceeding to such as are first, we should elevate ourselves from sensibles which have an ultimate subsistence. The soul, likewise, should for this purpose be liberated from all vice, in consequence of hastening to *the* (vision of the) *good*; and should ascend to the principle which is in herself, and become one instead of many things, in order that she may survey the principle of all things, and *the one*. Hence it is requisite, that the soul of him who ascends to *the good* should then become intellect, and that he should commit his soul to, and establish it in intellect, in order, that what intellect sees, his soul may vigilantly receive, and may through intellect survey *the one*; not employing any one of the senses, nor receiving any thing from them, but with a pure intellect, and with the summit (and as it were, flower) of intellect, beholding that which is most pure. When, therefore, he who applies himself to the survey of a thing of this kind, imagines that there is either magnitude, or figure, or bulk about this nature, he has not intellect for the leader of the vision; because intellect is not naturally adapted to perceive things of this kind, but such an energy is the energy of sense, and of opinion following sense. But in order to perceive *the one*, it is necessary to receive from intellect a declaration of what intellect is able to accomplish. Intellect, however, is able to see either things prior to itself, or things pertaining to itself, or things effected by itself. And the things contained in itself, are pure; but those prior to itself are still purer and more simple; or rather this must be asserted of that which is prior to it. Hence, that which is prior to it, is not intellect, but something more excellent. For intellect is a *certain* one among the number of being; but that is not a *certain* one, but is prior to everything. Nor is it being for being has, as it were, the form of *the one*. But that is formless, and is even without intelligible form. For the nature of *the one* being generative of all things, is not any one of them. Neither, therefore, is it a certain thing, nor a quality, nor a quantity, nor intellect nor soul, nor that which is moved nor again that which stands still. Nor is it in place, or in time; but is by itself uniform, or rather without form, being prior to all form, to motion and to permanency. For these subsist about being which also cause it to be multitudinous. Why, however, if it is not moved, does it not stand still? Because it is necessary that one or both of these should subsist about being. And that which stands still, stands still through permanency, and is not the same with it. Hence permanency is accidental to it, and it no longer remains simple. For when we say that the one is the cause of all things, we do not predicate anything as an accident to it, but rather as something which happens to us, because we possess something from it, *the one* in the meantime subsisting in itself. It is necessary, however, when speaking accurately of *the one*, neither to call it *that*, nor *this*. . . .

How does intelligence see; what does it see; and, in short, how does it subsist; and how is it generated from *the one*, so that it may see? For now indeed the soul perceives the necessity of the existence of these things. It desires, however, to understand this which is so much spoken of by the wise men of antiquity, viz. how from *the one* being such as we have said it is, each thing has its subsistence, whether it be multitude, or the duad, or number; and why *the one* did not abide in itself, but so great a multitude flowed from it, as is seen to have an existence, and which we think should be referred to *the one*. We must say, therefore, as follows, *invoking God himself, not with external speech, but with the soul itself, extending ourselves in prayer to him, since we shall then be able to pray to him properly, when we approach by ourselves alone to the alone.* It is necessary, therefore, that the beholder of him, being in himself as if in the interior part of a temple, and quietly abiding in an eminence beyond all things, should survey the statues as it were which are established outwardly, or rather that statue which first shines forth to the view, and after the following manner behold that which is naturally adapted to be beheld. With respect to everything that is moved, it is necessary there should be something to which it is moved. For if there is nothing of this kind, we should not admit that it is moved. But if any thing is generated posterior to that which the movable nature tends, it is necessary that it should always be generated in consequence of that prior cause being converted to itself. Let, however, the generation which is in time be now removed from us who are discoursing about eternal beings. And if in the course of the discussion we attribute generation to things which exist eternally, let it be considered as indicative of cause and order. Hence, that which is from thence generated, must be said to be generated, the cause not being moved. For if something was generated in consequence of that cause being moved, the thing generated after the motion would be the third, and not the second from the cause. It is necessary, therefore, the cause being immovable, that if any thing secondary subsists after it, this second nature should be produced, without the cause either verging to it, or consulting, or in short being moved. How, therefore, and what is it necessary to conceive about that abiding cause? We must conceive a surrounding splendour, proceeding indeed from this cause, but from it in a permanent state, like a light from the sun shining, and as it were running around it, and being generated from it, the cause itself always abiding in the same immovable condition. All beings, likewise, as long as they remain, necessarily produce from their own essence, about themselves, and externally from the power which is present with them, a nature whose hypostasis is suspended from them, and which is as it were an image of the archetype from which it proceeded. Thus fire emits from itself indeed heat, and snow not only retains cold within itself (but imparts it to other things). This, however, such things as are fragrant especially testify. For as long as they exist, something proceeds from them, of which whatever is near them partakes. All such things, likewise, as are now perfect generate; but that which is always perfect, always generates, and that which it produces is perpetual. It also generates something less than itself. What, therefore, is it requisite to say of that which is most perfect? Shall we say that nothing proceeds from it; or rather that the greatest things posterior to it are its progeny? But the greatest thing posterior to it, and the second, is intellect. For intellect sees it, and is in want of it alone. But this most perfect nature is not in want of intellect. It is also necessary that the thing generated from that which is better than intellect, should be intellect. And intellect is superior to all things after the first, because other things are posterior to it. Thus for instance, soul is the reason of intellect, and a certain energy of it, just as intellect of that first God (who is beyond intellect). But the reason of soul is indeed obscure. For as it is the image of intellect, on this account it is necessary that it should look to intellect. After the same manner also, it is necessary that intellect should look to the highest God, in order that it may be intellect. It sees him, however, not separated from him, but because it is after him, and there is nothing between; as neither is there anything between soul and intellect. But everything desires its generator. This also it loves, and especially when that which is generated and the generator are alone. When, however, that which generates is the most excellent of things, the thing begotten is necessarily present with it in such a manner, as to be separated by *otherness* alone.

But we say that intellect is the image of this most excellent nature. For it is necessary to speak more clearly. In the first place, indeed, it is necessary that intellect should in a certain respect be generated, and preserve (in itself) much of its generator; and also that it should have such a similitude to it, as light has to the sun. Its generator, however, is not intellect. How therefore did he generate intellect (so far as it is intellect)? May we not say, because intellect, by conversion, looks to him? But the vision itself is intellect. For that which apprehends another thing, is either sense or intellect. And sense indeed may be compared to a line but the other gnostic powers of the soul to a circle. A circle, however, of this kind is as it were partible. But this is not the case with intellect. Or may we not say that this also is one? But the one here is the power of all things. Hence intelligence surveys those things of which it is the power, divided as it were from the power; for otherwise it would not be intellect. For intellect now possesses from itself a co-sensation as it were of the great extent of its power; in which power, its essence, consists. Intellect, therefore, through itself defines its own being, by a power derived from him (i.e. from the first God), and perceives that essence is as it were one of the parts of and from him, and that it is corroborated by him, and perfected by and from him into essence. It sees, however, itself derived from thence, as something which is as it were partible from that which is impartible; and not only itself, but life, and intellection, and all things, because the first God is nothing of all things. For on this account all things are from him, because he is not detained by a certain form. For he is one alone. And intellect, indeed, in the order of beings is all things. But he on this account is none of the things which are in intellect; and all things which have a subsistence among beings are derived from him. Hence also these are essences. For they are now definite, and each possesses as it were a form. Being, however, ought not to be surveyed in that which is as it were indefinite, but as fixed by bound and permanency. But permanency in intelligibles is circumscription and form, in which also they receive their hypostasis. This intellect, therefore, which deserves the appellation of the most pure intellect, and which is of the genus of intelligibles, originates from no other source than the first principle. And being now generated, it generates together with itself beings, all the beauty of ideas, and all the intelligible Gods. . . .

On this account all things are distributed by Plato in a triple order about the king of all. For he says, "that all things are about the king of all; second things about that which is second and such as are third about that which ranks as the third." He also says that this king is father of cause, denominating intellect cause. For with Plato, intellect is the demiurgus. But he says that this cause produced soul in that *Crater* (mentioned by him in the *Timaeus*). The cause, however, being intellect, he says that the father is *the good*, and that which is beyond intellect, and beyond essence. In many places, also, he calls being and intellect idea; so that from Plato we may know that intellect and idea are from *the good*, but soul from intellect. . . .

It has been shown, however, as far as it is possible to demonstrate about things of this kind, that it is requisite to think that beyond being there is *the one*, such as reason wishes to unfold; that next to this, being and intellect subsist; and that, in the third place, follows the nature of soul. . . .

CHAPTER 14

ST. AUGUSTINE

AURELIUS AUGUSTINE WAS BORN on November 13, 354 A.D. at Tagaste in Numidia (modern Algeria). He was born of what may be termed a pagan father and a Christian mother. His father, however, was well versed in the various Greek philosophies prevalent at the time. Augustine was well educated and led what the early Christians termed an immoral life, which, in reality, was the common life of the time and would not be considered immoral in the broader sense of the word.

He was not inclined toward the Christian views of his mother, considering Christianity an extreme form of asceticism. During his extensive travels he first gave thought to metaphysics when he became associated with the philosophic school known as the Manichaeans. It later appeared to him too extreme and impractical in its doctrines and he abandoned the study. His thoughts were led to the higher life through the study of Platonic philosophy introduced to him by the neo-Platonic exponents of the time. He was eventually led to the embracing of Christianity by St. Ambrose.

After three years spent in retirement as head of a small monastic society, he was called to be a presbyter at Hippo and later became bishop of Hippo. He was opposed to the philosophical and religious doctrine that there were two separate worlds—a world of evil and a world of good. He declared that everything was of the good and he accounted for the evil conduct of man by the fact that man was given will and through the agency of will had the right to choose whether he desired to continue in the state of goodness or deviate from it. It is by his doctrine of will that he made his greatest contribution to Christianity and to philosophy.

St. Augustine was more than merely a theologian—he was a philosopher. Students of mysticism, philosophy, and Christianity alike will profit by the reading of his works entitled, *The City of God* and *The Confessions.*

ON FOREKNOWLEDGE AND FREE WILL

The manner in which Cicero addresses himself to the task of refuting the Stoics, shows that he did not think he could effect anything against them in the argument unless he had first demolished divination. And this he attempts to accomplish by denying that there is any knowledge of future things, and maintains with all his might that there is no such knowledge either in God or man, and that there is no prediction of events. Thus he both denies the foreknowledge of God, and attempts by vain arguments, and by opposing to himself certain oracles very easy to be refuted, to overthrow all prophecy, even such as is clearer than the light (though even these oracles are not refuted by him).

But in refuting these conjectures of the mathematicians, his argument is triumphant, because truly these are such as destroy and refute themselves. Nevertheless, they are far more tolerable who assert the fatal influence of the stars than they who deny the foreknowledge of future events. For, to confess that God exists, and at the same time to deny that He has foreknowledge of future things, is the most manifest folly. This Cicero himself saw, and therefore attempted to assert the doctrine embodied in the words of Scripture, "The fool hath said in his heart, There is no God." That, however, he did not do in his own person, for he saw how odious and offensive such an opinion would be; and therefore, in his book on the nature of the gods, he makes Cotta dispute concerning this against the Stoics, and preferred to give his own opinion in favor of Lucillius Balbus, to whom he assigned the defence of Stoical position, rather than in favor of Cotta, who maintained that no divinity exists. However, in his books on divination, he in his own person most openly opposes the doctrine of the prescience of future things. But all this he seems to do in order that he may not grant the doctrine of fate, and by so doing destroy free will. For he thinks that, the knowledge of future things being once conceded, fate follows as so necessary a consequence that it cannot be denied.

But, let these perplexing debatings and disputations of the philosophers go on as they may, we, in order that we may confess the most high and true God Himself, do confess His will, supreme power, and prescience. Neither let us be afraid lest, after all, we do not do by will that which we do by will, because He, whose foreknowledge is infallible, foreknew that we would do it. It is this which Cicero was afraid of, and therefore opposed foreknowledge. The Stoics also maintained that all things do not come to pass by necessity, although they contended that all things happen according to destiny. What is it then, that Cicero feared in the prescience of future things? Doubtless it was this,—that if all future things have been foreknown, they will happen in the order in which they have been foreknown; and if they come to pass in this order, there is a certain order of things foreknown by God; and if a certain order of things, then a certain order of causes, for nothing can happen which is not preceded by some efficient cause. But if there is a certain order of causes according to which everything happens which does happen, then by fate, says he, all things happen which do happen. But if this be so, then is there nothing in our own power, and there is no such thing as freedom of will; and if we grant that, says he, the whole economy of human life is subverted. In vain are laws enacted. In vain are reproaches, praises, chidings, exhortations had recourse to; and there is no justice whatever in the appointment of rewards for the good and punishments for the wicked. And that consequences so disgraceful, and absurd, and pernicious to humanity may not follow, Cicero chooses to reject the foreknowledge of future things, and shuts up the religious mind to this alternative, to make choice between two things, either that something is in our own power, or that there is foreknowledge,—both of which cannot be true; but if the one is affirmed, the other is thereby denied. He therefore like a truly great and wise man, and one who consulted very much and very skillfully for the good of humanity, of those two chose the freedom of the will, to confirm which he denied the foreknowledge of future things; and thus, wishing to make men free, he makes them sacrilegious. But the religious mind chooses both, and maintains both by the faith of piety. But how so? says Cicero; for the knowledge of future things being granted, there follows a chain of consequences which ends in this, that there can be nothing depending on our own free wills. And further, if there is anything depending on our wills, we must go backwards by the same steps of reasoning till we arrive at the conclusion that there is no foreknowledge of future things. For we go backwards through all the steps in the following order:—if there is free will, all things do not happen according to fate; if all things do not happen according to fate, there is not a certain order of causes; neither is there a certain order of things foreknown by God,—for things cannot come to pass except they are preceded by efficient causes,—but, if there is no fixed and certain order of causes foreknown by God, all things cannot be said to happen according as He foreknew that they would happen. And further, if it is not true that all things happen just as they have been foreknown by Him, there is not, says he, in God any foreknowledge of future events.

Now, against the sacrilegious and impious darings of reason, we assert both that God knows all things before they come to pass, and that we do by our free will whatsoever we know and feel to be done by us only because we will it. But that all things come to pass by fate, we do not say; nay we affirm that nothing comes to pass by fate; for we demonstrate that the name of fate, as it is wont to be used by those who speak of fate; meaning thereby the position of the stars at the time of each one's conception or birth, is an unmeaning word, for astrology itself is a delusion. But an order of causes in which the highest efficiency is attributed to the will of God, we neither deny nor do we designate it by the name of fate, unless, perhaps, we may understand fate to mean that which is spoken, deriving it from *fari*, to speak; for we cannot deny that it is written in the sacred Scriptures, "God hath spoken once; these two things have I heard, that power belongeth unto God. Also unto Thee, O God, belongeth mercy: For Thou wilt render unto every man according to his works." Now the expression, "Once hath He spoken," is to be understood as meaning "*immovably*," that is, unchangeably all things which shall be, and all things which He will do. We might, then, use the word fate in the sense it bears when derived from *fari*, to speak, had it not already come to be understood in another sense, into which I am unwilling that the hearts of men should unconsciously slide. But it does not follow that, though there is for God a certain order of all causes, there must therefore be nothing depending on the free exercise of our own wills, for our wills themselves are included in that order of causes which is certain to God, and is embraced by His foreknowledge, for human wills are also causes of human actions and He who foreknew all the causes of things would certainly among those causes not have been ignorant of our wills. For even that very concession which Cicero himself makes is enough to refute him in this argument. For what does it help him to say that nothing takes place without a cause, but that every cause is not fatal, there being a fortuitous cause, a natural cause, and a voluntary cause? It is sufficient that he confesses that whatever happens must be preceded by a cause. For we say that those causes which are called fortuitous are not a mere name for the absence of causes, but are only latent, and we attribute them either to the will of the true God, or to that of spirits of some kind or other. And as to natural causes, we by no means separate them from the will of Him who is the author and framer of all nature. But now as to voluntary causes. They are referable either to God, or to angels, or to men, or to animals of whatever description, if indeed those instinctive movements of animals devoid of reason, by which, in accordance with their own nature, they seek or shun various things, are to be called wills. And when I speak of the wills of angels, I mean either the wills of good angels, whom we call the angels of God, or of the wicked angels, whom we call the angels of the devil, or demons. Also by the wills of men I mean the wills either of the good or of the wicked. And from this we conclude that there are no efficient causes of all things which come to pass unless voluntary causes, that is, such as belong to that nature which is the spirit of life. The spirit of life, therefore, which quickens all things, and is the creator of every body, and of every created spirit, is God Himself, the uncreated spirit. In His supreme will resides the power which acts on the wills of all created spirits, helping the good, judging the evil, controlling all, granting power to some, not granting it to others. For, as He is the creator of all natures, so also is He the bestower of all powers, not of all wills; for wicked wills are not from Him, being contrary to nature, which is from Him. As to bodies, they are more subject to wills; some to our wills, by which I mean the wills of all living mortal creatures, but more to the wills of men than of beasts. But all of them are most of all subject to the will of God, to whom all wills are subject, since they have no power except what He has bestowed upon them. The cause of things, therefore, which makes but is not made is God; but all causes both make and are made. Such are all created spirits, and especially the rational. Material causes, therefore, which may rather be said to be made than to make, are not to be reckoned among efficient causes, because they can only do what the wills of spirits do by them. How, then, does an order of causes which is certain to the foreknowledge of God necessitate that there should be nothing which is dependent on our wills, when our wills themselves have a very important place in the order of causes? Cicero, then, contends with those who call this order of causes fatal, or rather designate this order itself by the name of fate; to which we have an abhorrence, especially on account of the word, which men have become accustomed to understand as meaning what is not true. But, whereas he denies that the order of all causes is most certain, and perfectly clear to the prescience of God, we detest his opinion more than the Stoics do. For he either denies that God exists,—which, indeed, in an assumed personage, he has labored to do, in his book, *De Natura Deorum*,—or if he confesses that He exists, but denies that He is prescient of future things, what is that but just "the fool saying in his heart there is no God." For one who is not prescient of all future things is not God. Wherefore our wills also have just so much power as God willed and foreknew that they should have; and therefore whatever power they have, they have it within most certain limits; and whatever they are to do, they are most assuredly to do, for He whose foreknowledge is infallible foreknew that they would have the power to do it, and would do it. Wherefore, if I should choose to apply the name of fate to anything at all, I should rather say that fate belongs to the weaker of two parties, will to the stronger, who has the other in his power, than that the freedom of our will is excluded by that order of causes, which, by an unusual application of the word peculiar to themselves, the Stoics call FATE.

CHAPTER 15

MOHAMMED

MOHAMMED WAS BORN AT Mecca in 570 A.D. In his twenty-fifth year he became manager of the estate of the widow Kadijah, and their ensuing marriage was a long and happy one. In the peaceful years which passed between his youth and his Illumination—at the age of forty—he retired yearly for a period of solitude and silence and prepared himself through meditation and fasting for the coming of Cosmic Consciousness.

Kadijah was the first to believe in the revelations Mohammed received, and, at first, only her faith stood between him and the mockery and disbelief which are the portion of all prophets. It is related that, after three unsuccessful years, he invited forty of his chief kindred to a banquet, told them of his mission, and asked who would second him. Only his sixteen-year-old cousin Ali responded. The magnitude of the enterprise to be undertaken by an old man and a youth appeared ridiculous to the rest, and the meeting dissolved in laughter. For thirteen years he traveled, preaching the revelations which were vouchsafed to him from time to time (and gathered in the *Koran*). Pursued by enemies, threatened with death, homeless and scorned, he finally arrived at Medina and there gained some adherents. He was now fifty-three, aging and desolate—except for the comfort from within—and he turned to defending his beliefs with the sword, backed by an increasing number of converts, so that after ten years of grueling activity the faith was rapidly spreading.

The tenets of the *Koran* bettered the social, as well as the spiritual and religious, conditions of the time and served to stimulate and enlighten the Arabian world so that, in the following centuries, it became a leader in civilization. Mystical students find the life of Mohammed an impressive illustration of Cosmic Attunement. Here was an unschooled man, far from the centers of civilized life, who yet laid the foundations of a social and spiritual regime destined to influence millions of persons throughout the centuries and up to the present moment!

In the essay, *Mohammed and Mohammedanism* Carlyle gives a sympathetic picture of the prophet and his mission. ". . . A spontaneous, passionate, yet just, true-meaning man! Full of wild faculty, fire and light; of wild worth, all uncultured; working out his life-task in the depths of the Desert there. . . ."

We are presenting three excerpts from the *Koran* as translated by E.H. Palmer. The first, from the Chapter of the Table, illustrates Mohammed's interest in the Master Jesus whom he mentions so frequently. The second, from the Chapter of the Night, should be compared with the Commandments in the *Bible*. The third, the Chapter of the Folding Up, presents a passage of poetic beauty.

THE KORAN

When God said, 'O Jesus, son of Mary! remember my favours towards thee and towards thy mother, when I aided thee with the Holy Ghost, till thou didst speak to men in the cradle and when grown up.

'And when I taught thee the Book and wisdom and the law and the gospel; when thou didst create of clay, as it were, the likeness of a bird, by my power, and didst blow thereon, it became a bird; and thou didst heal the blind from birth, and the leprous by my permission; and when thou didst bring forth the dead by my permission; and when I did ward off the children of Israel from thee, when thou didst come to them with manifest signs, and those who misbelieved amongst them said, "This is naught but obvious magic." . . .'

And when God said, 'O Jesus, son of Mary! is it thou who didst say to men, take me and my mother for two gods, beside God?' He said, 'I celebrate Thy praise! what ails me that I should say what I have no right to? If I had said it, Thou wouldst have known it; Thou knowest what is in my soul, but I know not what is in Thy soul; verily, Thou art one who knoweth the unseen. I never told them save what Thou didst bid me,—"Worship God, my Lord and your Lord," and I was a witness against them so long as I was amongst them; but when Thou didst take me away to thyself Thou wert the watcher over them, for Thou art witness over all. If Thou shouldst punish them, verily, they are Thy servants; if Thou shouldst forgive them, verily, Thou art the mighty and the wise.' God said, 'This is the day when their confession shall profit the confessors, for them are gardens beneath which rivers flow, to dwell therein for ever and for aye.'

God is well pleased with them, and they well pleased with Him; that is the mighty happiness.

God's is the kingdom of the heavens, and the earth, and all that is therein, and He is mighty over all.

II

Put not with God other gods, or thou wilt sit despised and forsaken.

Thy Lord has decreed that ye shall not serve other than Him; and kindness to one's parents, whether one or both of them reach old age with thee; and say not to them, 'Fie!' and do not grumble at them, but speak to them a generous speech. And lower to them the wing of humility out of compassion, and say, 'O Lord! have compassion on them as they brought me up when I was little!' Your Lord knows best what is in your souls if ye be righteous, and, verily, He is forgiving unto those who come back penitent.

And give thy kinsman his due and the poor and the son of the road; and waste not wastefully, for the wasteful were ever the devil's brothers; and the devil is ever ungrateful to his Lord.

But if thou dost turn away from them to seek after mercy from thy Lord, which thou hopest for, then speak to them an easy speech.

Make not thy hand fettered to thy neck, nor yet spread it out quite open, lest thou shouldst have to sit down blamed and straitened in means. Verily, thy Lord spreads out provisions to whomsoever He will or He doles it out. Verily, He is ever well aware of and sees His servants.

And slay not your children for fear of poverty; we will provide for them; beware! for to slay them is ever a great sin!

And draw not near to fornication; verily, it is ever an abomination, and evil is the way thereof.

And slay not the soul that God has forbidden you, except for just cause; for he who is slain unjustly we have given his next of kin authority; yet let him not exceed in slaying; verily, he is ever helped.

And draw not near to the wealth of the orphan, save to improve it, until he reaches the age of puberty, and fulfil your compacts; verily, a compact is ever enquired of.

And give full measure when ye measure out, and weigh with a right balance; that is better and a fairer determination.

And do not pursue that of which thou hast no knowledge; verily, the hearing, the sight, and the heart, all of these shall be enquired of.

And walk not on the earth proudly; verily, thou canst not cleave the earth, and thou shalt not reach the mountains in height.

III

In the name of the merciful and compassionate God.
When the sun is folded up,
And when the stars do fall,
And when the mountains are moved,
And when the she-camels ten months' gone with young shall be neglected,
And when the beasts shall be crowded together,
And when the seas shall surge up,
And when souls shall be paired with bodies,
And when the child who was buried alive shall be asked for what sin she was slain,
And when the pages shall be spread out,
And when the heaven shall be flayed,
And when hell shall be set ablaze,
And when Paradise shall be brought nigh,
The soul shall know what it has produced!
I need not swear by the stars that slink back, moving swiftly, slinking into their dens!
Nor by the night when darkness draws on!
Nor by the morn when it first breathes up!
Verily, it is the speech of a noble apostle, mighty, standing sure with the Lord of the throne, obeyed and trusty too!
Your comrade is not mad; he saw him on the plain horizon, nor does he grudge to communicate the unseen.

Nor is it the speech of a pelted devil.
Then wither do ye go?
It is but a reminder to the worlds, to whomsoever of you pleases to go straight:—but ye will not please, except God, the Lord of the world, should please.

CHAPTER 16

AVICENNA

AVICENNA, AS ABU ALI al-Husayn Ibn Abdallah Ibn Sina is known to Europe, was born in 979 A.D. in the province of Bukhara, Persia. He was one of the greatest of Arabian (Islamic) physicians and philosophers. At the age of ten he was well acquainted with the Koran and the Arabic classics, and for the following six years he studied philosophy, mathematics, astronomy, and medicine. In connection with his philosophical studies he memorized the *Metaphysics* of Aristotle, but its meaning was not understood by him until he chanced upon the commentary of Alfarabi, who was the compiler of the first encyclopedia of Rosicrucian science and art. At the age of seventeen Avicenna proved his medical knowledge by curing the Samanid ruler, Nuh Ibn Mansur, from a dangerous illness, the event leading to a post in court and access to the royal library.

After his father's death, Avicenna traveled quite extensively, and lectured for a period of time on logic and astronomy at Jurjan, near the Caspian. Later, when in Hamadan, he gained the favor of Shams Addaula and was made vizier; but the soldiery mutinied against their sovereign and demanded the life of the new vizier. Under the successor of Shams Addaula, Avicenna was imprisoned in a fortress, but managed to escape and fled to Isfahan. When the ruler of Isfahan captured Hamadan in 1024 A.D., Avicenna spent the remaining thirteen years of his life as court physician to Ala Addaula, as well as general literary and scientific adviser.

Avicenna led a very arduous life, and although he traveled a great deal, he still found time to write—many of his writings contributing greatly to our Rosicrucian literature and teachings. About 100 treatises are ascribed to Avicenna, but his most influential work was his *Canon of Medicine*, which was translated into many different languages and used as a textbook in both the Orient and the Occident.

Avicenna was greatly influenced in his philosophical studies by Aristotle and to some degree by Neo-Platonic ideas. It was his desire to reconcile philosophy and religion, an object which was the endeavor of many later Arabic and Jewish philosophers. However great his attainments in the field of metaphysics, logic, and astronomy, the eminence which Avicenna gained lies chiefly in his contributions to philosophy and medicine not only in his own country but to the advancement of philosophy and medicine throughout the world.

ON MEDICINE

Medicine considers the human body as to the means by which it is cured and by which it is driven away from health. The knowledge of anything, since all things have causes, is not acquired or complete unless it is known by its causes. Therefore in medicine we ought to know the causes of sickness and health. And because health and sickness and their causes are sometimes manifest, and sometimes hidden and not to be comprehended except by the study of symptoms, we must also study the symptoms of health and disease. Now it is established in the sciences that no knowledge is acquired save through the study of its causes and beginnings, if it has had causes and beginnings; nor completed except by knowledge of its accidents and accompanying essentials. Of these causes there are four kinds: material, efficient, formal, and final.

Material causes, on which health and sickness depend, are—the affected member, which is the immediate subject, and the humors; and in these are the elements. And these two are subjects that, according to their mixing together, alter. In the composition and alteration of the substance which is thus composed, a certain unity is attained.

Efficient causes are the causes changing and preserving the conditions of the human body; as airs, and what are united with them; and evacuation and retention; and districts and cities, and habitable places, and what are united with them; and changes in age and diversities in it, and in races and arts and manners, and bodily and animate movings and restings, and sleeping and wakings on account of them; and in things which befall the human body when they touch it, and are either in accordance or at variance with nature.

Formal causes are physical constitutions, and combinations and virtues which result from them.

Final causes are operations. And in the science of operations lies the science of virtues, as we have set forth. These are the subjects of the doctrine of medicine; whence one inquires concerning the disease and curing of the human body. One ought to attain perfection in this research; namely, how health may be preserved and sickness cured. And the causes of this kind are rules of eating and drinking, and the choice of air, and the measure of exercise and rest; and doctoring with medicines and doctoring with the hands. All this with physicians is according to three species: the well, the sick, and the medium of whom we have spoken.

CHAPTER 17

ST. ANSELM

SAINT ANSELM, ARCHBISHOP OF Canterbury, was born in Aosta, Italy, in 1033. From childhood he was fired with the love of learning, but upon reaching the age of fifteen—after having been refused admittance to a monastery—he was diverted by the pleasures of youth and his inherent ardour for learning was temporarily lost. Like Abelard, he led a wandering life through France, as was the custom of the scholars of those days.

In 1060 he entered the monastery of Bec, in Normandy, where he studied under the illustrious Lanfranc. In three years he became prior, and in 1078, abbot of this monastery, which under his guidance became famous as a center of learning. Meanwhile, Lanfranc had become Archbishop of Canterbury, and when he died in 1089, the king, William Rufus, seized the revenues of the see, made no new appointment, and for four years kept the Church of England in a state of anarchy. Such were the conditions that existed when the king, in a moment of repentance when he thought he was dying, refused to consider Anselm's protests and appointed the well-loved Abbot of Bec to this high position in 1092.

Thus began Anselm's tumultuous career as Archbishop of Canterbury, with its many embroilments with William Rufus and his successor, Henry I. However, his indomitable spirit—even when subjected to banishment—proved his resoluteness of character, and as a stalwart champion of the Church, he became one of the chief figures in religious history. As a writer and thinker he may claim yet higher rank, for it is not often that a Catholic saint wins the admiration of German philosophers and English historians. In 1720, Clement XI placed him in the list of Church authorities, his works being recognized as a pattern for all theologians. He died April 21, 1109, and the day of his death is observed in the Roman Catholic Church.

Anselm's chief achievement in philosophy was the ontological argument for the existence of God put forth in his *Proslogion,* wherein he strove to demonstrate the existence of God from the conception of a perfect thing.

PROSLOGION, OR A DISCOURSE ON THE BEING OF GOD

This good thou art, thou, God the Father; this is thy Word, that is, thy Son. For nothing, other than what thou art, or greater or less than thou, can be in the Word by which thou dost express thyself; for thy Word is true as thou art truthful. And hence it is truth itself, just as thou art; no other truth than thou; and thou art of so simple a nature, that of thee nothing can be born other than which thou art. This very good is the one love common to thee and to thy Son; that is, the Holy Spirit proceeding from both. For this love is not unequal to Thee or to thy Son; seeing that thou dost love thyself and him, and he, thee, and himself, to the whole extent of thy being and his. Nor is there aught else proceeding from thee and from him, which is not unequal to thee and to him. Nor can anything proceed from the supreme simplicity, other than what this, from which it proceeds, is.

But what each is, separately, this is all the Trinity at once, Father, Son, and Holy Spirit; seeing that each separately is none other than the supremely simple unity, and the supremely unitary simplicity, which can neither be multiplied nor varied. Moreover, there is a single necessary Being. Now, this is that single necessary Being, in which is every good; nay, which is every good, and a single entire good, and the only good.

And now, my soul, arouse and lift up all thy understanding, and conceive, so far as thou canst, of what character and how great is that good. For, if individual goods are delectable, conceived in earnestness how delectable is that good which contains the pleasantness of all goods; and not such as we have experienced in created objects but as different as the Creator from the creature. For, if the created life is good, how good is the creative life! If the salvation given is delightful, how delightful is the salvation which has given all salvation! If wisdom in the knowledge of the created world is lovely, how lovely is the wisdom which has created all things from nothing! Finally, if there are many great delights in delectable things, what and how great is the delight in him who has made these delectable things!

Who shall enjoy this good? And what shall belong to him, and what shall not belong to him? At any rate, whatever he shall wish shall be his, and whatever he shall not wish shall not be his. For, these goods of body and soul will be such as eye hath not seen nor ear heard, neither has the heart of man conceived (Isaiah lxiv. 4; 1 Corinthians ii. 9).

Why, then, dost thou wander abroad, slight man, in thy search for the goods of thy soul and body? Love the one good in which are all goods, and it sufficeth. Desire the simple good which is every good, and it is enough. For, what dost thou love, my flesh? What dost thou desire, my soul? There, there is whatever ye love, whatever ye desire.

If beauty delights thee, there shall the righteous shine forth as the sun (Matthew xiii. 43). If swiftness or endurance, or freedom of body, which naught can withstand, delight thee, they shall be as angels of God,—because it is shown a natural body: it is raised a spiritual body (1 Corinthians xv. 44)—in power certainly, though not in nature. If it is a long and sound life that pleases thee, there a healthful eternity is, and an eternal health. For the righteous shall live forever (Wisdom v. 15), and the salvation of the righteous is of the Lord (Psalms xxxvii. 39). If it is satisfaction of hunger, they shall be satisfied when the glory of the Lord hath appeared (Psalms xxviii. 15). If it is quenching a thirst, they shall be abundantly satisfied with the fatness of thy house (Psalms xxxvi. 8). If it is melody, there the choirs of angels sing forever, before God. If it is any not impure, but pure pleasure, thou shalt make them drink of the river of thy pleasures, O God (Psalms xxxvi. 8).

If it is wisdom that delights thee, the very wisdom of God will reveal itself to them. If friendship, they shall love God more than themselves, and one another as themselves. And God shall love them more than they themselves; for they love him, and themselves, and one another, through him, and he, himself and them, through himself. If concord, they all have a single will.

If power, they shall have all power to fulfil their will, as God to fulfil his. For, as God will have power to do what he wills, through himself, so they will have power, through him, to do what they will. For, as they will not will aught else than he, he shall will whatever they will; and what he shall will cannot fail to be. If honor and riches, God shall make his good and faithful servants rulers over many things (Luke xii. 42); nay, they shall be called sons of God, and gods; and where his Son shall be, there they shall be also, heirs indeed of God, and joint-heirs with Christ (Romans viii. 17).

If true security delights thee, undoubtedly they shall be as sure that those goods, or rather that good, will never and in no wise fail them; as they shall be sure that they will not lose it of their own accord; and that God, who loves them, will not take it away from those who love him against their will; and that nothing more powerful than God will separate him from them against his will and theirs.

But what, or how great, is the joy, where such and so great is the good! Heart of man, needy-heart, heart acquainted with sorrows, nay, overwhelmed with sorrows, how greatly wouldst thou rejoice, if thou didst abound in all those things! Ask thy inmost mind whether it could contain its joy over so great a blessedness of its own.

Yet assuredly, if any other whom thou didst love altogether as thyself possessed the same blessedness, thy joy would be doubled, because thou wouldst rejoice not less for him than for thyself. But, if two, or three, or many more, had the same joy, thou wouldst rejoice as much for each one as for thyself, if thou didst love each as thyself. Hence, in that perfect love of innumerable blessed angels and sainted men, where none shall love another less than himself, every one shall rejoice for each of the others as for himself.

If, then, the heart of man will scarce contain his joy over his own so great good, how shall it contain so many and so great joys? And doubtless, seeing that every one loves another so far as he rejoices in the other's good, and as, in that perfect felicity, each one should love God beyond compare, more than himself and all the others with him; so he will rejoice beyond reckoning in the felicity of God more than in his own and that of all the others with him.

But if they shall so love God with all their heart, and all their mind, and all their soul, that still all the heart, and all the mind, and all the soul shall not suffice for the worthiness of this love; doubtless they will so rejoice with all their heart, and all their mind, and all their soul, that all the heart, and all the mind, and all the soul shall not suffice for the fulness of their joy.

My God and my Lord, my hope and the joy of my heart, speak unto my soul and tell me whether this is the joy of which thou tellest us through thy Son: Ask and ye shall receive, that your joy may be full (John xvi. 24). For I have found a joy that is full, and more than full. For when heart, and mind, and soul, and all the man, are full of that joy, joy beyond measure will still remain. Hence, not all of that joy shall enter into those who rejoice; but they who rejoice shall wholly enter into that joy.

Show me, O Lord, show thy servant in his heart whether this is the joy into which thy servants shall enter, who shall enter into the joy of their Lord. But that joy, surely, with which thy chosen ones shall rejoice, eye hath not seen nor ear heard, neither has it entered into the heart of man (Isaiah lxiv. 4; 1 Corinthians ii. 9). Not yet, then, have I told or conceived, O Lord, how greatly those blessed ones of thine shall rejoice. Doubtless they shall rejoice according as they shall love; and they shall love according as they shall know. How far they will know thee, Lord, then! and how much they will love thee! Truly, eye hath not seen, nor ear heard, neither has it entered into the heart of man in this life, how far they shall know thee and how much they shall love thee in that life.

I pray, O God, to know thee, to love thee, that I may rejoice in thee. And if I cannot attain to full joy in this life may I at least advance from day to day, until that joy shall come to the full. Let the knowledge of thee advance in me here, and there be made full. Let the love of thee increase, and there let it be full, that here my joy may be great in hope, and there full in truth. Lord, through thy Son thou dost command, nay, thou dost counsel us to ask; and thou dost promise that we may receive, that our joy may be full. I ask, O Lord, as thou dost counsel through our Wonderful Counsellor. I will receive what thou dost promise by virtue of thy truth, that my joy may be full. Faithful God, I ask I will receive, that my joy may be full. Meanwhile, let my mind meditate upon it; let my tongue speak of it. Let my heart love it; let my mouth talk of it. Let my soul hunger for it: let my flesh thirst for it; let my whole being desire it, until I enter into thy joy, O Lord, who art the Three and the One God, blessed for ever and ever. Amen.

CHAPTER 18

MOSES MAIMONIDES

MOSES BEN MAIMON, BETTER known under the Greek form of his name, Maimonides, was the outstanding intellectual of medieval Judaism.

During the Arab era of civilization, very little new thought was contributed to the ancient philosophies, and most intellectual effort was devoted to eclectic practices, that is, the combining of the outstanding principles, doctrines, and tenets of the different philosophies into one, a sort of a tendency toward unification. The greatest work of Maimonides was his combining of the old Jewish theology with the Aristotelian philosophy. He was also recognized as an outstanding authority and commentator on the Old Testament and the Talmud.

He was born in 1135 A.D. at Cordova, Spain. Mohammedan fanatics from Africa, invading Cordova in 1148, drove his family out. It was thought that, for a period of approximately eleven years, he had been converted to Mahommedanism, embracing the Islam faith at Fez. This is much doubted, however. He spent years wandering, finally settling in Cairo. There, during a period of thirty years, he became first a jeweler, then later a physician, and finally received the distinction of being appointed court physician. He died in 1204 A.D.

The rationality of his mind is to be particularly noted in the way he devised a method to approach the problem of proving the existence of God. As was the custom of the time, and his experience, he takes the better elements of thought and logic from the old philosophies and from the prevailing religions, using them as a means to arrive at his final conclusions.

If all students of the philosophies and metaphysics and all seekers for truth, and those who profess to have a sincere desire to fathom life's mysteries, would approach the same and other problems as broadly and as fairly as this ancient thinker did, a much clearer conception of life would be had and realities would not be cast aside because of prejudice or intolerance but would be weighed for their true worth, regardless of their source.

METHOD FOR PROVING GOD'S EXISTENCE

My proof of God's existence, as far as I now can explain it in general terms, is as follows. The universe is either eternal or has had a beginning; if it had a beginning, there must necessarily exist a being which caused the beginning; this is clear to common sense; for a thing that has had a beginning, cannot be the cause of its own beginning. Another being must have caused it.

The universe was, therefore, created by God. If on the other hand the universe were eternal, it could in various ways be proved that, apart from the things which constitute the universe, there exists a being which is neither body nor a force in a body, and which is one, eternal, not preceded by any cause, and immutable. That being is God. You see that the proofs for the Existence, the Unity and the Incorporeality of God must vary according to the propositions admitted by us. Only in this way we can succeed in obtaining a perfect proof, whether we assume the eternity or the creation of the universe. For this reason you will find in my works on the Talmud, whenever I have to speak of the fundamental principles of our religion, or to prove the existence of God, that I employ arguments which imply the eternity of the universe. I do not believe in that eternity, but I wish to establish the principle of the existence of God by an indisputable proof, and should not like to see this most important principle founded on a basis which every one could shake or attempt to demolish, and which others might consider as not being established at all; especially when I see that the proofs of the philosophers are based on those visible properties of things, which can only be ignored by persons possessing certain preconceived notions, while the Mutakallemim establish their arguments on propositions which are to such an extent contrary to the actual state of things as to compel these arguers to deny altogether the existence of the laws of nature. When I shall have to treat of the creation, I shall in a special chapter prove my opinion to some extent, and shall attain the same end which every one of the Mutakallemim had in view, yet I shall not contradict the laws of nature, or reject any such part of the Aristotelian theory as has been proved to be correct. Even the most cogent of the proofs offered by the Mutakallemim respecting the act of creation, has only been obtained by reversing the whole order of things and by rejecting everything fully demonstrated by the philosophers. I, however, shall be able to give a similar proof without ignoring the laws of nature and without being forced to contradict facts which have been clearly perceived. I find it necessary to mention to you the general propositions of the Mutakallemim, by which they prove the act of creation, the existence of God, His Unity and His Incorporeality. I intend to explain their method, and also to point out the inferences which are to be drawn from each proposition. After this, I shall describe those theories of the philosophers which are closely connected with our subject, and I shall then explain their method.

Do not ask me to prove in this work the propositions of the philosophers, which I shall briefly mention to you; they form the principal part of Physics and Metaphysics. Nor must you expect that I should repeat the arguments of the Mutakallemim in support of their propositions, with which they wasted their time, with which the time of future generations will likewise be wasted, and on which numerous books have been written. Their propositions, with few exceptions, are contradicted by the visible properties of things, and beset with numerous objections. For this reason they were obliged to write many books and controversial works in defense of their theories, for the refutation of objections, and for the reconciliation of all apparent contradictions, although in reality this object cannot be attained by any sophistical contrivance. As to the propositions of the philosophers which I shall briefly explain, and which are indispensable for the demonstration of the three principles—the Existence, the Unity, and the Incorporeality of God, they will for the greater part be admitted by you as soon as you shall hear them and understand their meaning; whilst in the discussion of other parts reference must be made for their proofs to works on Physics and Metaphysics, and if you direct your attention to such passages as will be pointed out to you, you will find everything verified that requires verification.

I have already told you that nothing exists except God and this universe, and that there is no other evidence for His Existence but this universe in its entirety and in its several parts. Consequently the universe must be examined as it is; the propositions must be derived from those properties of the universe which are clearly perceived, and hence you must know its visible form and its nature. Then only will you find in the universe evidence for the existence of a being not included therein. I have considered it, therefore, necessary to discuss first in a merely colloquial manner, in the next chapter, the totality of existing things, and to confine our remarks to such as have been fully proved and established beyond all doubt. In subsequent chapters I shall treat of the propositions of the Mutakallemim, and describe the method by which they explain the four fundamental principles. In the chapters which will follow, I propose to expound the propositions of the philosophers and the methods applied by them in verifying those principles. In the last place, I shall explain to you the method applied by me in proving those four principles, as I have stated to you.

CHAPTER 19

ROGER BACON

ROGER BACON WAS BORN in England in 1214. He was educated at Oxford. After graduating from Oxford, he went to the University of Paris to study. While there he became concerned with the ancient knowledge and sciences preserved by the Arabs, and he enjoyed spending more hours in discoursing on science than in discussing theology with the theologians at the university. At that time there was very little scientific spirit in the land due to the church's suppression of the ancient sciences and also due to the church's great concern with theological dogma and creed. He came back to England in about 1250 and spent much time in study and experimentation. He joined the Franciscan Order. In 1257 the Order interdicted any publication of his work.

Roger Bacon was a deep student of mysticism and of occultism. Many of his writings are allegorical. The symbols which he used to signify certain fundamental forces of the universe have led many a superficial student of his life to think him a devotee of Black Magic. They failed to realize that it became necessary because of the persecution of those discoursing on mysticism, occultism, and science, in his time, to veil their subject matter. Later his books were condemned for "suspected novelties" by Jerome de Ascoli, the general of the Order. He was thrown into prison and remained there for a time. He died in 1294. His greatest work was the *Opus Majus* which is considered an encyclopedia of knowledge of his time. It contained two primary ideas of importance—the first is that in order to be a science, a subject must be sufficiently understood to be mathematically stated; the second is that experience is all important in scientific investigation.

Roger Bacon's name has been associated with the invention and perfection of gunpowder. He also wrote of cars of the future that would be self-propelled and of boats that would be propelled by a force of some sort and be free of oars and sails. From his revelations he wrote of devices which even today might seem fantastic, but which in our progressive era we cannot exactly set aside as an improbability, for even those things which we have not even conceived of and which were written about by Bacon are more probable than the other things which he wrote about and which in his time seemed fantastic but which have come to pass.

In contributing to the knowledge of humanity, he depended upon two sources for his wisdom—scientific investigation of the mysteries of the universe and Divine revelation. He gives equal credit for the knowledge that was his to these two sources. It is because of his dependence upon revelations, or what we might term "Cosmic intuition," that many of the later materialists, scientists, depending upon pure investigation and research, through their objective senses, caused the biographies ridiculing Bacon. Today, when the limitations of the senses are reached, rather than admit that a thing does not exist, there is a tendency toward resorting to a metaphysical inquiry into the subject. The following is an excerpt from Bacon's works on experimental science.

ON EXPERIMENTAL SCIENCE

Having laid down the main points of the wisdom of the Latins as regards language, mathematics and optics, I wish now to review the principles of wisdom from the point of view of experimental science, because without experiment it is impossible to know anything thoroughly.

There are two ways of acquiring knowledge, one through reason, the other by experiment. Argument reaches a conclusion and compels us to admit it, but it neither makes us certain nor so annihilates doubt that the mind rests calm in the intuition of truth, unless it finds this certitude by way of experience. Thus many have arguments toward attainable facts, but because they have not experienced them, they overlook them and neither avoid a harmful nor follow a beneficial course. Even if a man that has never seen fire, proves by good reasoning that fire burns, and devours and destroys things, nevertheless the mind of one hearing his arguments would never be convinced, nor would he avoid fire until he puts his hand or some combustible thing into it in order to prove by experiment what the argument taught. But after the fact of combustion is experienced, the mind is satisfied and lies calm in the certainty of truth. Hence argument is not enough, but experience is.

This is evident even in mathematics, where demonstration is the surest. The mind of a man that receives that clearest of demonstrations concerning the equilateral triangle without experiment will never stick to the conclusion nor act upon it till confirmed by experiment by means of the intersection of two circles from either section of which two lines are drawn to the ends of a given line.

Then one receives the conclusion without doubt. What Aristotle says of the demonstration by the syllogism being able to give knowledge, can be understood if it is accompanied by experience, but not of the bare demonstration. What he says in the first book of the Metaphysics, that those knowing the reason and cause are wiser than the experienced, he speaks concerning the experienced who know the bare fact only without the cause. But I speak here of the experienced that know the reason and cause through their experience. And such are in their knowledge, as Aristotle wishes to be in the sixth book of the Ethics, whose simple statements are to be believed as if they carried demonstration as he says in that very place.

Whoever wishes without proof to revel in the truths of things need only know how to neglect experience. . . .

Experience is of two kinds. One is through the external senses: such are the experiments that are made upon the heaven through instruments in regard to facts there, and the facts on earth that we prove in various ways to be certain in our own sight. And facts that are not true in places where we are, we know through other wise men that have experienced them. Thus Aristotle with the authority of Alexander, sent 2,000 men throughout various parts of the earth in order to learn at first hand everything on the surface of the world, as Pliny says in his Natural History. And this experience is human and philosophical just as far as man is able to make use of the beneficent grace given to him, but such experience is not enough for man, because it does not give full certainty as regards corporeal things because of their complexity and touches the spiritual not at all. Hence man's intellect must be aided in another way, and thus the patriarchs and prophets who first gave science to the world secured inner light and did not rest entirely on the senses. So also many of the faithful since Christ. For grace makes many things clear to the faithful, and there is divine inspiration not alone concerning spiritual but even about corporeal things. In accordance with which Ptolemy says in the Centilogium that there is a double way of coming to the knowledge of things, one through the experiments of science, the other through divine inspiration, which latter is far the better as he says.

Of this inner experience there are seven degrees, one through spiritual illumination in regard to scientific things. The second grade consists of virtue, for evil is ignorance as Aristotle says in the second book of the Ethics. And Algazel says in the logic that the mind is disturbed by faults, just as a rusty mirror in which the images of things cannot be clearly seen, but the mind is prepared by virtue like a well polished mirror in which the images of things show clearly.... And this is our experience, because a known truth draws men into its light for love of it, but the proof of this love is the sight of the result. And indeed he that is busy against truth must necessarily ignore this, that it is permitted him to know how to fashion many high sounding words and to write sentences not his own, just as the brute that imitates the human voice or an ape that attempts to carry out the works of men, although he does not understand their purpose. Virtue, then clears the mind so that one can better understand not only ethical, but even scientific things....

The third degree of spiritual experience is the gift of the Holy Spirit, which Isaiah describes. The fourth lies in the beatitudes which our Lord enumerates in the Gospels. The fifth is the spiritual sensibility. The sixth is in such fruits as the peace of God, which passeth all understanding. The seventh lies in states of rapture and in the methods of those also, various ones of whom receive it in various ways, that they may see many things which it is not permitted to speak of to man. And whoever is thoroughly practised in these experiences or in many of them, is able to assure himself and others, not only concerning spiritual things, but all human knowledge. And indeed, since all speculative thought proceeds through arguments which either proceed through a proposition by authority or through other propositions of argument, in accordance with this which I am now investigating, there is a science that is necessary to us, which is called experimental. I wish to explain this, not only as useful to philosophy, but to the knowledge of God and the understanding of the world: as in a former book I followed language and science to their end, which is the Divine wisdom by which all things are ordered.

CHAPTER 20

JAN VAN RUYSBROECK

JAN VAN RUYSBROECK WAS born in 1293 in a small village of the same name near Brussels. He was raised by his uncle, a canon of the Cathedral of St. Gudule, and by another devout priest. In 1317 he took orders, and for twenty-six years thereafter served as a Cathedral chaplain—a position obtained through his uncle's influence. During this period he passed through the mystical stages described in the first two books of *The Adornment of the Spiritual Marriage*, so that by the time he was fifty it seemed that he could no longer endure the contrast between religious formalism, as exemplified by the Cathedral life, and his own growing spiritual intuitions. So in 1343, with the two men who had raised him, he left Brussels forever to devote himself to a life of prayer and contemplation. The authorities gave them the old hermitage of Groenendael in the forest of Soignes, and there they were soon joined by disciples and formed a small community.

Although Ruysbroeck devoted much of his time to meditation, thus receiving the truths expressed in his books, yet he believed in the life of "balanced action" and sought to be of service to all his fellows in the daily life of the priory and to advise and enlighten the ever-increasing number of disciples who journeyed to Groenendael because of him.

Ruysbroeck's books show that, in the ordinary course of life, he accepted all the restrictions imposed by the church and state of the day, never questioning practices which were definitely wrong—from a mystical point of view—or endeavoring to reform either church or state. But in the hours of inspiration he transcended these misleading boundaries, charted out by temporal authorities, and followed the light of the God Within, so that the greatest part of his work is timeless, and as useful a guide to The Way as it was in the fourteenth century.

Aside from *The Adornment of the Spiritual Marriage*, his best known early works are *The Kingdom of God's Lovers, The Mirror of Eternal Salvation, The Seven Cloisters*, and *The Seven Degrees of Love*. However, three later books, containing the essence of his earlier works, are considered his finest contributions to the eternal search. These are: *The Twelve Beguines, The Book of Supreme Truth*, and *The Sparkling Stone*. From the last of these we have chosen excerpts which describe the differences between three states through which the seeker must pass: as a "faithful servant," a "secret friend," and a "hidden son" of God.

HIDDEN SONS OF GOD

If, further, this good man would become an inward and ghostly man, he needs must have three further things. The first is a heart unencumbered with images; the second is spiritual freedom in his desires, the third is the feeling of inward union with God. . . . Whosoever then has, in his inward exercise, an imageless and free ascent unto his God, and means nought else but the glory of God, must taste of the goodness of God; and he must feel from within a true union with God. And in this union, the inward and spiritual life is made perfect; for in this union, the desirous power is perpetually enticed anew and stirred to new inward activity. And by each act, the spirit rises upwards to a new union. . . .

Further, you must know that if this ghostly man would now become a God-seeing man, he needs must have three other things. The first is the feeling that the foundation of his being is abysmal, and he should possess it in this manner; the second is that his inward exercise should be wayless; the third is his indwelling should be a divine fruition.

Now understand, you who would live in the spirit, for I am speaking to no one else. The union with God which a spiritual man feels, when the union is revealed to the spirit as being abysmal—that is, measureless depth, measureless height, measureless length and measureless breadth—in this manifestation the spirit perceives that through love it has plunged itself into the depth and has ascended into the height and escaped into the length; and it feels itself to be wandering in the breadth, and to dwell in a knowledge which is ignorance. And through this intimate feeling of union, it feels itself to be melting into the Unity; and through dying to all things, into the life of God. And there it feels itself to be one life with God. And this is the foundation, and the first point, of the God-seeing life.

And from this there arises the second point, which is an exercise above reason and without condition: for the Divine Unity, of which every God-seeing spirit has entered into possession in love, eternally draws and invites the Divine Persons and all loving spirits into itself. And this inward drawing is felt by each lover, more or less, according to the measure of his love and the manner of his exercise. . . . And therefore we must all found our lives upon a fathomless abyss; that we may eternally plunge into Love, and sink down in the fathomless Depth. And with that same Love, we shall ascend, and transcend ourselves, in the incomprehensible Height. And in that Love which is wayless, we shall wander and stray, and it shall lead us and lose us in the immeasurable Breadth of the Love of God. And herein we shall flee forth and flee out of ourselves, into the unknown raptures of the Goodness and Riches of God. And therein we shall melt and be melted away, and shall eternally wander and sojourn within the Glory of God. Behold! by each of these images, I show forth to God-seeing men their being and their exercise, but none else can understand them. For the contemplative life cannot be taught. But where the Eternal Truth reveals Itself within the spirit all that is needful is taught and learnt. . . .

We must now observe the great difference which there is between the faithful servants and the inward friends of God. . . . All such friends God calls and invites inwards, and He teaches them the distinctions of inward exercises and many a hidden way of ghostly life. But He sends His servants outwards, that they may be faithful to Him and to His House in every service and in every kind of outward good works.

Behold, thus God gives His grace and His help to each man according to his fitness; that is, according to the way in which he is in tune with God, whether in outward good works or in the inward practice of love. But none can do and feel the inward exercises unless he be wholly turned inward to God. For as long as a man is divided of heart, so long he looks outwards, and is unstable of mind, and is easily swayed by joy and grief in temporal things, for these are still alive within him. And though he may live according to the commandments of God, inwardly he abides in darkness, and knows not what inward exercises may be, nor how these should be practiced. . . .

That One Thing which is needful for all men is Divine love. The better part is an inward life, with loving adherence to God. This Mary Magdalen had chosen and this is chosen by the secret friends of God. But Martha chose an outward, unenclosed, and active life; and that is the other part, in which one may serve God, but which is neither so perfect nor so good. . . .

But further we find a more subtle and inward difference, between the secret friends and the hidden sons of God; and yet both these alike by their inward exercise maintain themselves in the Presence of God. But the friends possess their inwardness as an attribute, for they choose the loving adherence to God as best and highest of all that they ever can and will reach: and that is why they cannot with themselves and their own activity penetrate to the imageless Nudity. For they have, as images and intermediaries between God and themselves, their own being and their own activity. And though in their loving adherence they feel united with God, yet, in this union, they always feel a difference and an otherness between God and themselves. For the simple passing into the Bare and Wayless, they do not know and love: and therefore their highest inward life ever remains in Reason and in Ways. And though they have clear understanding and discernment of all virtues that may be conceived, the simple staring with open heart into the divine Brightness remains unknown of them. And though they feel themselves uplifted to God in a mighty fire of love, yet they keep something of their own selfhood, and are not consumed and burnt to nothingness in the unity of love.... Nevertheless, you should know that all good and faithful men are the sons of God; for they are all born of the Spirit of God, and the Spirit of God lives in them. And he moves and stirs them—each according to his own capacity—to virtues and good works, wherein they are well pleasing to God. But because of the inequality of their adherence and their exercises, I call some the faithful servants of God, and others I call His secret friends, and others again His hidden sons. . . .

If we are born of the Spirit of God, we are the sons of grace; and so our whole life is adorned with virtues.... In this birth all good men are sons of God. And the spirit of God kindles and stirs each one of them in particular to those virtues and to those good works for which he is in readiness, and of which he is capable. And so they please God all in common, and each in particular, according to the measure of his love and the nobleness of his exercise; nevertheless, they do not feel established nor possessed of God, nor assured of eternal life for they may still turn away and fall into sin. And that is why I call them rather servants and friends, than sons. But when we transcend ourselves, and become, in our ascent towards God, so simple that the naked love in the height can lay hold of us, where love enfolds love, above every exercise of virtue—that is, in our Origin, of Which we are spiritually born—then we cease, and we and all our selfhood die in God. And in this death we become hidden sons of God, and find a new life within us: and that is eternal life....

When we go towards God by means of the virtues, God dwells in us; but when we go out from ourselves and from all else, then we dwell in God. So soon as we have faith, hope and charity, we have received God, and He dwells in us with His grace, and He sends us out as His faithful servants, to keep His commandments. And He calls us in again as His secret friends, so soon as we are willing to follow His counsels; and He names us openly as His sons so soon as we live in opposition to the world. But if above all things we would taste God, and feel eternal life in ourselves, we must go forth into God with our feeling, above reason; and there we must abide, onefold, empty of ourselves, and free from images, lifted up by love into the simple bareness of our intelligence. For when we go out in love beyond and above all things, and die to all observation in ignorance and in darkness, then we are wrought and transformed through the Eternal Word, Who is the Image of the Father. In this idleness of our spirit, we receive the Incomprehensible Light, which enwraps us and penetrates us, as the air is penetrated by the light of the sun. And this Light is nothing else than a fathomless staring and seeing. What we are, that we behold; and what we behold, that we are: for our thought, our life, and our being are uplifted in simplicity, and made one with the Truth which is God.

CHAPTER 21

ST. THOMAS AQUINAS

THOMAS AQUINAS LIVED AT the very beginning of the era of the renaissance of knowledge, the time when learning was being restored. He was born in 1225 in a little town outside of Naples, Italy. At the age of seventeen he joined the Dominican Order and became a student of the other famous scholastic philosopher, Albertus Magnus of Cologne. His duties necessarily made him take an active part in the various church affairs, but he did an enormous amount of work in theology and philosophy.

He was a typical scholastic, well learned in Latin and in Greek, and an ardent student and admirer of Aristotle. He attempted to establish a new system of knowledge by combining the scientific methods of Aristotle with the theological ones of the church. He recognized two different forms of knowledge.

First was Divine Revelation, inspirations, or, as we would say, Cosmic Attunement, wherein man receives an influx into his consciousness of Infinite Wisdom which expresses itself outwardly. The other form of knowledge was that expounded by Aristotle—observation of the facts of reality through the senses and the classification of them by the reason. He considered, however, that both of these forms of knowledge had their source in God.

We may see from this that St. Thomas Aquinas was a true metaphysician. He died March 7, 1274. Below we bring to you excerpts from his writings on the theory of knowledge.

THEORY OF KNOWLEDGE

There are certain intellectual substances which are called immaterial.

The substances mentioned before that we called immaterial are necessarily also intellectual. Anything, indeed, is intellectual which is free from matter because of its very nature *to be known* can mean only *to be perceived*, for to be actually intelligible and an intellect in activity are the same things.

It is clear, moreover, that anything is actually intelligible when it is separated from matter, for we cannot have intellectual knowledge of material things except by abstracting it from the matter of the thing. Hence the same judgment must be given concerning the intellect, that what are immaterial are also intellectual. For example, the immaterial substances are the first and highest in rank among individual things, for actuality naturally precedes potentiality. Moreover, the intellect appears superior to all other things, for the intellect uses bodies as instruments. Immaterial substances must, therefore, be intellectual. To this purpose as much as some things among individual things are of higher grade, so much the more do they approach to the likeness of God. We see things of the lowest grade to participate in the divine nature only so far as to exist, although inanimate; that some things, such as plants, both exist and live; that some, such as animals, even know things; but that the highest grade of existence and the one approaching the nearest to God is in the understanding. Hence the highest creatures are intellectual and because of their approach nearer to the likeness of God than any other creature they are said to be made in the likeness of God.

The substance through which man knows is the lowest in order of intellectual substances.

Since, moreover, it is not possible in the case of things to proceed to an infinite number of orders, just as there was found the highest substance among those previously mentioned which approached nearest to God, so there is necessarily found a lowest one which approaches nearest to corporeal matter, and this indeed can be made plain. For *to know* places man above the other animals. It is clear that man alone understands general truths and the customs of things and immaterial things which are perceived only through intelligence. It is impossible, moreover, that *to know* is an act exercised through the bodily organism the same as sight is a function of the eye. It is necessary that the whole instrument of knowledge should be free from that kind of a thing which is known through itself; just as the pupil is free from colors of its own. So, indeed, colors are recognized insofar as the kinds of color are received in the pupil, but the receiving part must be free from that which is received. The intellect knows all natural things of sense. If, then, it knows through a bodily organ, that organ must be free from any natural sensible thing, which is impossible; for example, every reason knows things that can be known by the species (general truths) becoming known to itself, for this is its first principle of knowledge. The intellect, moreover, knows things in an immaterial way, even those things that are in their very nature material, by selecting the universal form (or truth) from the individual material conditions. It is impossible, therefore, that the genus of a thing known in the intellect is material. Therefore, it is not known through some bodily organ, for every bodily organ is material. In the same way it appears from this that the sense is weakened and destroyed by too much sensibility, just as hearing by great noises or sight by things too bright, which happens because the harmony of the organism is destroyed; but the intellect grows stronger by the excellence of the things it knows for he that knows the higher things is able not only to know other things but to know them better. So, therefore, since man is found to be intelligent, and knowledge does not come through a bodily organ, there must be some other incorporeal substance through which man knows. For since this is able to be done without the body, the essence of it does not depend upon the body. For all characteristics and forms which cannot exist of themselves without the body are not able to act except through the body. Thus heat does not become hot through itself, but the body becomes hot through heat. That incorporeal substance, therefore, through which man knows is the lowest in kind of intellectual substances and the nearest approach to matter.

Concerning the characteristic of the intellect and the process of knowledge.

Since, moreover, to be intelligent is a higher grade of existence than to be sensible, just as the intellect is higher than the senses; and since, moreover, the things lower in the order of being, imitate the higher, just as the corporeal things, subject to generation and destruction, imitate the cycle of the heavenly bodies; so things of sense must be assimilated in their own way to things intelligible, and so, from the likeness of sensory things (to us), we are able to arrive at the knowledge of intelligible things. There is, moreover, in sensibles something of the highest grade that is active, such as the form (plan), and something of the lowest grade that is only potential, which is matter; also something intermediate compounded of matter and form. So also in the intelligible nature, for the highest intellectuality, which is God, is pure activity. The other intellectual substances are those having something of the active and potential after the nature of an intelligible being. The lowest of the intellectual substances through which man knows is only in potentiality, like an intelligible being. This is witnessed by the fact that man is found in the beginning only potentially intelligent, and afterwards, little by little, is led to active intelligence—and thence that through which man knows is called a *positive intellect*.

That the possible intellect in man receives intelligible forms from sensory things.

Since it is true, as said, that by as much as the intellectual nature is the higher, it has more general intelligible ideas, it follows that the human intellect, which we called possible, has less general ideas than other intellectual substances, and thence it is that it receives intelligible forms from sensory things. This is also evident in another way, if one considers. For the form must be proportionate to the susceptibility. Just as the possible human intellect among intellectual substances is found nearer to the corporeal matter, so it is necessary that its forms be nearest to material things.

That man needs potential sensory things for the understanding.

We must consider, also, that forms (truths) in corporeal things are individual and have a material existence; that in the intellect they are general and immaterial; which indeed shows the way in which we learn. For we know things universally and immaterially. The operation of understanding through intelligible forms (general truths, etc.), by which we understand, necessarily follows. Since one cannot pass from extreme to extreme save through a medium, the forms from corporeal things must pass to the intellect through some medium. Of such a nature are the potential sensory substances (qualities) that receive the forms of material things without the matter. The form of the stone comes to the eye, but not the matter, yet the forms of things in particular are received in the potential sensory substances (qualities); for in these potential sensory qualities we know only particulars. Therefore, it is necessary for man, in order to learn, to have senses. The proof of this is that if a sense is lacking a man lacks all knowledge of the sensory facts which are understood through that sense, just as a man born blind has no idea whatever of color.

That it is necessary to presuppose an active intellect.

From the above it is plain that the knowledge of things in our intellect is not caused through participation, or by the influx, in the act of knowing, of certain forms existing in themselves, as Plato believed, and others following him, but the intellect acquires knowledge from sensible things through mediating senses. But since in potential sensory things the forms are particular, as has been said, they are not actively intelligent but only potentially intelligible. The mind, indeed, does not know universals, moreover, it exists potentially; it is not led to activity except by some other agent. Therefore, there must be some other agent which makes forms existing in potentially intelligible things to be actually intelligible. The human intellect cannot do this for it is itself more potentially intelligible than actively so. It is necessary then to presuppose another intellect which makes forms potentially intelligible to be actively so, just as light makes colors that are potentially visible be actually visible, and this we call the active intellect, which would not have to be supposed if forms were themselves actually intelligible, as the Platonists have said. Therefore, in knowing there must be first our possible intellect which receives the intelligible forms, and second, an active intellect which makes these forms actually intelligible. Since the intellect is perfected through intelligible forms it is called the intellect of use since it holds these intelligible forms so that it can use them as it wishes, midway between mere potentiality and complete activity. Since it also holds the general truths before mentioned in complete activity it is called the active intellect for it actively comprehends a thing when the class of the thing has been made a form of the active intellect. Therefore, it is said that the active intellect is knowledge and activity.

—Compendium Theologiae

CHAPTER 22

MEISTER ECKHART

JOHANNES ECKHART WAS BORN in Hochheim, Thuringia, in what is now East Germany. He joined the Dominican Order at Erfurt and rose to the position of Vicar. He has been called the father of the German mystics, also the philosophical, creative genius of the German mystics and the father of German speculation. The famous Eckhart manuscripts are recognized by the German Rosicrucians and by the Rosicrucian Order throughout the world as the work of a Rosicrucian. He was a learned member of the Dominican or preaching order and for some time *lector biblicus* at the University of Paris, also of the Dominican College of St. Jacob where he was given his title Meister by Pope Boniface VIII.

It is not exactly known, but it is thought that he graduated in Cologne in the Scholasticism of Albertus Magnus (1205-1281) and Thomas Aquinas (1225-1274) whose system was at that time rapidly acquiring its hold. He held different administrative and political posts in the province. He was well known because of the fact that he preferred teaching in what was termed the "vulgar tongue" in contrast to scholastic Latin. He gained a great number of followers among the masses at large and the laity. The growth of his following and popularity alarmed the church authorities. They accused him of numerous things, principally because of the fact that he preached in the "vulgar tongue" instead of Latin. He barely escaped being declared a heretic by the Church in 1329, after his death. The charges were numerous—in fact, he was charged with preaching seventeen different, heretical doctrines.

It is a pleasure to give the student in mysticism and philosophy as well as to the Rosicrucian, some of his writings, as presented by Franz Pfeiffer and translated by C. De B. Evans.

THE DIVINE BEING

No man can tell of God exactly what he is. According to St. Dionysius, God is not anything we can say or think. St. Augustine cried: "I who have ever been in God and ever more shall be, would sooner I had never been and never should be than that we found a single word that we could say of God. Were we compelled to speak of God, in that case I should say: Verily, in no sense is God comprehensible nor yet unattainable. God is what thought cannot better." Nay, I declare God beggars human thought; he transcends all human conception. No man knows what God is. Aught that a man could or would think of God, God is not at all. It is the nature of the soul not to be satisfied except with God. But all that heart can desire is small, is insignificant compared with God. Yet man's thought may be never so rich or so rare but his desire outstrips it. So he transcends man's desire as well as transcending human thought.

St. Dionysius says God is naught. Meaning that God is as incomprehensible as naught. St. Bernard says, I know not what God is; but what I know not that he is that same is he. A heathen philosopher maintains that what we know of the First Cause is rather what we are ourselves than what the First Cause is. For that passes understanding. And in this strain the heathen doctor argues in his book, "The Light of Lights," that God is super-essential, super-rational, super-intelligible, i.e., beyond the natural understanding. I speak not of gracious understanding. By grace man may be carried to the length of understanding as St. Paul understood who was caught up into the third heaven and saw unspeakable things. He saw, but was not able to express them. For what a man knows he knows in its cause or in its mode or in its effect. But in these respects God remains unknown, for he is the first. Further, he is modeless, i.e., undetermined. And he is without effect, that is, in his mysterious stillness. Here he abides apart from the names that are given him. Moses asked his name. God answered, He who is hath sent thee. Otherwise he could not tell it. God as simply being, in that sense he could never give himself to be known to creature. Not that he could not do it, but creature could not understand it.— I have often laid it down that God's lordship does not lie merely in his lordship over creatures; his lordship consists in his power to create a thousand worlds and dominate them all in his abstract essence. Therein lies his lordship. Dionysius and Gregory both teach that the Divine Being is not comprehensible in any sense; not to any wit nor any understanding, not even to angelic mind even at its clearest. It was said by a philosopher that whoso knows of God that he is unknown, that man knows God. For it is the height of gnosis and perception to know and understand in agnosia and a-perception. To know him really is to know him as unknowable. As the master puts it: If I must speak of God, then I will say, God is something which is in no sense to be reached or grasped; and I know nothing else about him. According to St. Augustine, what we say about God is not true;

what we say that God is he is not; what we say he is not that he is rather than what we say that he is. Nothing we can say of God is true. God's worth and God's perfection cannot be put into words. When I say man, I have in my mind human nature. When I say grey, I have in my mind the greyness of grey. When I say God, I have in my mind neither God's majesty nor his perfection. Dionysius insists that the more we can abstract from God the better by negation than by affirmation. Hence the dictum of one master that to argue about God from likeness is to argue falsely about him, but to argue by denials is to argue about him correctly. Dionysius says, writing about God, He is super-essential, he is super-luminous; he attributes to him neither this nor that. For whatever he conceives, God far transcends it. There is no knowing him by likeness. Rather by attributing unlikeness may we make some approach to understanding him. Take an illustration. Supposing I describe a ship to someone who has never seen one, then on looking at a stone he will plainly see that it is not a ship. And the plainer he sees that it is not ship-like, the more he will know about a ship. It is the same with God. The more we impute to him not-likeness, the nearer do we get to understanding him. Holy Scripture yields us merely privatives. That we should credit God with matter, form and work is due to our gross senses. We fail to find God one because we try to come at him by likeness. Dionysius cries: "Friend Timothy, if thou wouldst catch the spirit of truth pursue it not with the human senses. It is so swift, it comes rushing." God is to be sought in opposites; in knowing shall we know God; in forgetfulness of ourselves and all things even to the naked essence of the Godhead. Dionysius was exhorting one of his disciples. "Friend," quoth he, "cease from all activity and empty thyself of self that thou mayst commune with the Sovran Good, God, namely." Pray God we may seek him so that we shall find him nevermore to lose him. Amen.

Poverty

The really virtuous man does not want God. What I have I want not. He makes no plans, he sets no store by things. As God is higher than man, so is he readier to give than man is to receive. Not by his fasts and vigils and his many outward works does a man prove his progress in the virtuous life, but it is a sure sign of his growth if he finds eternal things more and more attractive than the things that pass. The man who has a thousand marks of gold and gives it all away for love of God is doing a fine thing; yet I say, it were far finer and far better for him to despise it, setting it at naught on God's account.

A man should orient his will and all his works to God and having only God in view go forward unafraid, not thinking, am I right or am I wrong. One who worked out all the chances ere starting his first fight would never fight at all. And if, going to some place, we must think how to set the front foot down we shall never get there. It is our duty to do the next thing; go straight on, that is the right way.

There are five kinds of poverty. The first is devilish poverty; the second, golden poverty; the third is willing poverty; the fourth is spiritual poverty; the fifth, divine poverty.

The first, or devilish poverty, applies to all who have not what they fain would have, outward or inward. That is their hell.

The second, golden poverty, is theirs who in the midst of goods and properties pass empty in and out. If everything they own was burnt the effect on them would be to leave them quite unmoved. Heaven must needs be theirs and they would have no less.

The third is willing poverty and belongs to those who, renouncing goods and honours, body and soul, leave everything with right good grace. These give judgment with the twelve apostles and by pronouncing judgment it is their judgment day who, knowing what they leave, yet set another in their heart and mightily bestir themselves about their own departures. Such are the willing poor.

The fourth are spiritual poor. These have forsaken friends and kindred, not merely goods and honour, body and soul; further they are quit of all good works; the eternal Word does all their work while they are idle and exempt from all activity. And since in the eternal Word is neither bad nor good, therefore they are absolutely empty.

The fifth are godly poor, for God can find no place in them to work in. Theirs is riddance without and within for they are bare and free from all contingent form. This is the man: in this man all men are one man and that man is Christ. Of him one master says, "Earth was never worthy of this man who looks on heaven and earth the same." This man is object-free in time and in eternity.

Not enough of those who have no object in eternity, but one thing more of those who are objectless in time. What is meant by object? There are two objects; one is otherness (not I); the other is a man's own proper self (his I).

The first otherness is becoming, all that has come into existence; such things breed otherness and pass away. This applies to the passage of time.

He who knows one matter in all things remains unmoved. For matter is the subject of form and there can be no matter without form nor form devoid of matter. Form without matter is nothing at all; but matter ever cleaves to form and is one undivided whole in every single part of it. Now, since form in itself is naught, therefore it moves nothing. And since matter is perfectly impartiable, therefore it is unmoved. This man then is unmoved by form or matter and is therefore objectless in time.

Man's other object is to possess his proper self, to identify himself with all perfection, with that most precious treasure his own aught: that is his quest. Now, when a thing has gotten its own form, no more nor less, that thing is all its own and no one else's. He who conceives this really is perfect in the sense that he is wholly objectless to eternity, etc.

CHAPTER 23

THOMAS Á KEMPIS

THOMAS HEMERKEN IS BETTER known as Thomas à Kempis. It is thought that he derived his nom de plume from his native town of Kempen. This little town was near the Rhine about forty miles north of Cologne. Thomas à Kempis was born in the year 1379 or 1380. He was a member of the Order of the Brothers of The Common Life which is understood to be one of the secret societies for the promulgation of mystic writings and practices of the Middle Ages. Many of these societies and organizations using names quite foreign to the Rosicrucian official name and title were, nevertheless, a part of the official body of the Rosicrucian Order.

Thomas à Kempis spent the last seventy years of his life at Mount St. Agnes, a famous monastery of Augustinian canons in the Diocese of Utrecht. Here he died on July 29, 1471, after a very active and eventful life. Most of the years of his life, while confined in the monastery, were devoted to copying ancient manuscripts with the object of perpetuating the work of teachers and thinkers of the past; reading and composing; and the peaceful routine of monastic piety. The famous work *Imitation of Christ* is perhaps one of the most popular Christian works ever witten and compares favorably in popularity with the Bible. Manuscripts of it in Latin are quite extensive throughout Europe and it was believed that the original was also written in Latin about the 15th Century. The exact date of composition is unknown. In fact, the authorship of the work is debated. Most authorities, however, recognize it as being of Thomas à Kempis. The work is not an original one. It consists of the interpretation of the writings and thoughts of the ancient mystics. It is really a compilation of medieval writings united by Thomas à Kempis into a spiritual work which could not have been done if it were not for his insight and his sincerity of purpose. It is with pleasure that we give to you excerpts from this famous book. We suggest that every reader attempt to secure a copy either from a bookseller or from the public library.

ON SPIRITUAL PROFIT

Open not thine heart to every man, but deal with one who is wise and feareth God. Be seldom with the young and with strangers. Be not a flatterer of the rich, nor willingly seek the society of the great. Let thy company be the humble and the simple, the devout and the gentle, and let thy discourse be concerning things which edify. Be not familiar with any woman, but commend all good women alike unto God. Choose for thy companions God and his Angels only, and flee from the notice of men.

We must love all men, but not make close companions of all. It sometimes falleth out that one who is unknown to us is highly regarded through good report of him, whose actual person is nevertheless unpleasing to those who behold it. We sometimes think to please others by our intimacy, and forthwith displease them the more by the faultiness of character which they perceive in us.

* * *

Avoid as far as thou canst the tumult of men; for talk concerning worldly things, though it be innocently undertaken is a hindrance, so quickly are we led captive and defiled by vanity. Many a time I wish that I had held my peace, and had not gone amongst men. But why do we talk and gossip so continually, seeing that we so rarely resume our silence without some hurt done to our conscience? We like talking so much because we hope by our conversations to gain some mutual comfort, and because we seek to refresh our wearied spirits by variety of thoughts. And we very willingly talk and think of those things which we love or desire or else of those which we most dislike.

But alas! It is often to no purpose and in vain. For this outward consolation is no small hindrance to the inner comfort which cometh from God. Therefore must we watch and pray that time pass not idly away. If it be right and desirable for thee to speak, speak things which are to edification. Evil custom and neglect of our real profit tend much to make us heedless of watching over our lips. Nevertheless, devout conversation on spiritual things helpeth not a little to spiritual progress, most of all where those of kindred mind and spirit find their ground of fellowship in God.

* * *

We may enjoy abundance of peace if we refrain from busying ourselves with the sayings and doings of others, and things which concern not ourselves. How can he abide long time in peace who occupieth himself with other men's matters, and with things without himself, and meanwhile payeth little or rare heed to the self within? Blessed are the single-hearted, for they shall have abundance of peace.

How came it to pass that many of the Saints were so perfect, so contemplative of Divine things? Because they steadfastly sought to mortify themselves from all wordly desires, and so were enabled to cling with their whole heart to God, and be free and at leisure for the thought of Him. We are too much occupied with our own affections, and too anxious about transitory things. Seldom, too, do we entirely conquer even a single fault, nor are we zealous for daily growth in grace. And so we remain lukewarm and unspiritual.

Were we fully watchful of ourselves, and not bound in spirit to outward things, then might we be wise unto salvation, and make progress in Divine contemplation. Our great and grievous stumbling-block is that, not being freed from our affections and desires, we strive not to enter into the perfect way of the Saints. And when even a little trouble befalleth us, too quickly are we cast down, and fly to the world to give us comfort.

If we would quit ourselves like men, and strive to stand firm in the battle, then should we see the Lord helping us from Heaven. For He Himself is always ready to help those who strive and who trust Him; yea, He provideth for us occasions of striving, to the end that we may win the victory. If we look upon our progress in religion as a progress only in outward observance and forms, our devoutness will soon come to an end. But let us lay the axe to the very root of our life, that, being cleansed from affections, we may possess our souls in peace.

If each year should see one fault rooted out from us, we should go quickly on to perfection. But on the contrary, we often feel that we were better and holier in the beginning of our conversion than after many years of profession. Zeal and progress ought to increase day by day; yet now it seemeth a great thing if one is able to retain some portion of his first ardour. If we would put some slight stress on ourselves at the beginning, then afterwards we should be able to do all things with ease and joy.

It is a hard thing to break through a habit, and a yet harder thing to do contrary to our own will. Yet if thou overcome not slight and easy obstacles, how shalt thou overcome greater ones? Withstand thy will at the beginning, and unlearn an evil habit, lest it lead thee little by little into worse difficulties. Oh, if thou knewest what peace to thyself thy holy life should bring to thyself, and what joy to others, methinketh thou wouldst be more zealous for spiritual profit.

CHAPTER 24

NICOLAUS COPERNICUS

MIKOLAJ KOPERNIK, POLISH SCHOLAR, best known by the Latinized form, Nicolaus Copernicus, was born February 19, 1473, in what was then Prussia. He entered the University of Cracow in 1491, and in 1495 he went to Padua where he studied medicine. In 1500 he was called to Rome and lectured in Mathemathics there. About 1507 he began to believe that the Earth revolved around the Sun. From that time on until his death he worked more or less intermittently on the exposition of his theory.

He aroused the ire of the church ecclesiasts by his statement that if it were true as he concluded, that the Earth was not the center of the universe as had been thought for centuries, and that there were numerous other planets equal in size to the size of the Earth, or perhaps even greater than the Earth, as he had reason to believe, then the Earth must not have been chosen alone for the theater of humanity. Man, then, if the Earth was not the center of the universe, would not be as important an element in the Cosmic Plan as he considered himself. He would not be the object of God's sole plan.

Copernicus was severely reprimanded by the church. Every effort was made to suppress his doctrine as well as his scientific teachings, because of this opposition to the church's dogma. Just as his book was about to be published he died, May 24, 1543. His life quickened the coming of the renaissance in learning. Below we give his arguments refuting the contention that the Earth is the center of the universe. They are worthy of your careful study from a historical and scientific point of view.

WHETHER THE EARTH HAS A CIRCULAR MOTION, AND CONCERNING THE LOCATION OF THE EARTH

As it has been already shown that the earth has the form of a sphere, we must consider whether a movement also coincides with this form, and what place the earth holds in the universe. Without this there will be no secure results to be obtained in regard to the heavenly phenomena. The great majority of authors of course agree that the earth stands still in the center of the universe, and consider it inconceivable and ridiculous to suppose the opposite.

But if the matter is carefully weighed it will be seen that the question is not yet settled and therefore by no means to be regarded lightly. Every change of place which is observed is due, namely, to a movement of the observed object or of the observer, or to movements of both, naturally in different directions, for if the observed object and the observer move in the same manner and in the same direction no movement will be seen. Now it is from the earth that the revolution of the heavens is observed and it is produced for our eyes. Therefore if the earth undergoes no movement this movement must take place in everything outside of the earth, but in the opposite direction than if everything on the earth moved, and of this kind is the daily revolution. So this appears to affect the whole universe, that is, everything outside the earth with the single exception of the earth itself. If, however, one should admit that this movement was not peculiar to the heavens, but that the earth revolved from west to east, and if this was carefully considered in regard to the apparent rising and setting of the sun, the moon and the stars, it would be discovered that this was the real situation. Since the sky, which contains and shelters all things, is the common seat of all things, it is not easy to understand why motion should not be ascribed rather to the thing contained than to the containing, to the located rather than to the location. From this supposition follows another question of no less importance, concerning the place of the earth, although it has been accepted and believed by almost all, that the earth occupies the middle of the universe. But if one should suppose that the earth is not at the center of the universe, that, however, the distance between the two is not great enough to be measured on the orbits of the fixed stars, but would be noticeable and perceptible on the orbit of the sun or of the planets: and if one was further of the opinion that the movements of the planets appeared to be irregular as if they were governed by a center other than the earth, then such an one could perhaps have given the true reasons for the apparently irregular movement. For since the planets appear now nearer and now father from the earth, this shows necessarily that the center of their revolutions is not the center of the earth: although it does not settle whether the earth increases and decreases the distance from them or they their distance from the earth.

Refutation of the Arguments of the Ancients That the Earth Remains Still In the Middle of the Universe, As If It Were Its Center

From this and similar reasons it is supposed that the earth rests at the center of the universe and that there is no doubt of the fact.

But if one believed that the earth revolved, he would certainly be of the opinion that this movement was natural and not arbitrary. For whatever is in accord with nature produces results which are the opposite of those produced by force. Things upon which force or an outside power has acted, must be injured and cannot long endure: what happens by nature, however, preserves itself well and exists in the best condition. So Ptolemy feared without good reason that the earth and all earthly objects subject to the revolution would be destroyed by the act of nature, since this latter is opposed to artificial acts, or to what is produced by the human spirit. But why did he not fear the same, and in a much higher degree, of the universe, whose motion must be as much more rapid as the heavens are greater than the earth? Or has the heaven become so immense because it has been driven outward from the center by the inconceivable power of the revolution; while if it stood still, on the contrary, it would collapse and fall together? But surely if this is the case the extent of the heavens would increase infinitely. For the more it is driven higher by the outward force of the movement, so much the more rapid will the movement become, because of the ever increasing circle which must be traversed in 24 hours; and conversely if the movement grows the immensity of the heavens grows. So the velocity would increase the size and the size would increase the velocity unendingly. According to the physical law that the endless cannot wear away nor in any way move, the heavens must necessarily stand still. But it is said that beyond the sky no body, no place, no vacant space, in fact nothing at all exists; then it is strange that some thing should be enclosed by nothing. But if the heaven is endless and is bounded only by the inner hollow, perhaps this establishes all the more clearly the fact that there is nothing outside the heavens, because everything is within it, but the heaven must then remain unmoved. The highest proof on which one supports the finite character of the universe is its movement. But whether the universe is endless or limited we will leave to the physiologues; this remains sure for us that the earth enclosed between the poles, is bounded by a spherical surface. Why therefore should we not take the position of ascribing to a movement conformable to its nature and corresponding to its form, rather than suppose that the whole universe whose limits are not and cannot be known moves? and why will we not recognize that the appearance of a daily revolution belongs to the heavens, but the actuality to the earth; and that the relation is similar to that of which one says: "We run out of the harbor, the lands and cities retreat from us." Because if a ship sails along quietly, everything outside of it appears to those on board as if it moved with the motion of the boat, and the boatman thinks that the boat with all on board is standing still, this same thing may hold without doubt of the motion of the earth, and it may seem as if the whole universe revolved. What shall we say, however, of the clouds and other things floating, falling or rising in the air—except that not only does the earth move with the watery elements belonging with it, but also a large part of the atmosphere, and whatever else is in any way connected with the earth; whether it is because the air immediately touching the earth has the same nature as the earth, or that the motion has become imparted to the atmosphere. A like astonishment must be felt if that highest region of the air be supposed to follow the heavenly motion, as shown by those suddenly appearing stars which the Greeks call comets or bearded stars, which belong to that region and which rise and set like other stars. We may suppose that part of the atmosphere, because of its great distance from the earth, has become free from the earthly motion. So the atmosphere which lies close to the earth and all things floating in it would appear to remain still, unless driven here and there by the wind or some other outside force, which chance may bring into play; for how is the wind in the air different from the current in the sea? We must admit that the motion of things rising and falling in the air is in relation to the universe a double one, being always made up of a rectilinear and a circular movement. Since that which seeks of its own weight to fall is essentially earthy, so there is no doubt that these follow the same natural law as their whole; and it results from the same principle that those things which pertain to fire are forcibly driven on high. Earthly fire is nourished with earthly stuff, and it is said that the flame is only burning smoke. But the peculiarity of the fire consists in this that it expands whatever it seizes upon, and it carries this out so consistently that it can in no way and by no machinery be prevented from breaking its bonds and completing its work. The expanding motion, however, is directed from the center outward; therefore if any earthly material is ignited it moves upward. So to each single body belongs a single motion, and this is evinced preferably in a circular direction as long as the single body remains in its natural place and its entirety. In this position the movement is the circular movement which as far as the body itself is concerned is as if it did not occur. The rectilinear motion, however, seizes upon those bodies which have wandered or have been driven from their natural position or have been in any way disturbed. Nothing is so much opposed to the order and form of the world as the displacement of one of its parts. Rectilinear motion takes place only when objects are not properly related, and are not complete according to their nature because they have separated from their whole and have lost their unity. Moreover, objects which have been driven outward or away, leaving out of consideration the circular motion, do not obey a single, simple and regular motion, since they cannot be controlled simply by their lightness or by the force of their weight, and if in falling they have at first a slow movement the rapidity of the motion increases as they fall, while in the case of earthly fire which is forced upwards—and we have no means of knowing any other kind of fire—we will see that its motion is slow as if its earthly origin thereby showed itself. The circular motion, on the other hand, is always regular, because it is not subject to an intermittent cause. Those other objects, however, would cease to be either light or heavy in respect to their natural movement if they reached their own place, and thus they would fit into that movement. Therefore, if the circular movement is to be ascribed to the universe as a whole and the rectilinear to the parts, we might say that the revolution is to the straight line as the natural state is to sickness. That Aristotle divided motion into three sorts, that from the center out, that inward toward center, and that around about the center, appears to be merely a logical convenience, just as we distinguish point, line and surface, although one cannot exist without the others, and none of them are found apart from bodies. This fact is also to be considered, that the condition of immovability is held to be nobler and diviner than that of change and inconstancy, which latter therefore should be ascribed rather to the earth than to the universe, and I would add also that it seems inconsistent to attribute motion to the containing and locating element rather than to the contained and located object, which the earth is. Finally since the planets plainly are at one time nearer and at another time farther from the earth, it would follow, on the theory that the universe revolves, that the movement of the one and same body which is known to take place about a center, that is the center of the earth, must also be directed toward the center from without and from the center outward. The movement about the center must therefore be made more general, and it suffices if that single movement be about its own center. So it appears from all these considerations that the movement of the earth is more probable than its fixity, especially in regard to the daily revolution, which is most peculiar to the earth.

CHAPTER 25

PARACELSUS

PHILLIPUS AUREOLUS THEOPHRASTUS BOMBAST von Hohenheim, known as Paracelsus, the student and scientist, was born November 10 (or 14), 1493; experienced transition September 24, 1541. Many of the fundamentals of alchemy, surgery, and medicine were taught him by his father. He studied further with the monks in the convent of St. Andrew. At the age of sixteen he entered the University of Basel, Switzerland. He traveled extensively throughout the East learning much of the Oriental philosophies and literature. He was initiated in the mystery schools of the Orient. He later became professor of medicine in the University of Basel. He taught, however, in an unorthodox manner, that is, he did not subscribe to the customs and manner of teaching of his adopted science. He became known as a prominent healer and was accorded recognition throughout all of Europe. He made enemies of his colleagues because he went beyond the limits of the prescribed medical knowledge and used some of the wisdom gained from his association with the Rosicrucians and mystery schools of the East. He was attacked from all angles—his morals, his concepts, and even physically. The world today, however, is beginning to appreciate his contribution to medicine and pharmacy. Rosicrucian students, however, have long appreciated his knowledge of the infinite laws and inner workings of man.

TRUE FAITH

Regarding the true and the false faith, Paracelsus says: "It is not a faith in the existence of a historical Jesus Christ that has the power to save mankind from evil, but a faith in the Supreme Power (God), through which the man Jesus was enabled to act. The former 'faith' is merely a belief and a result of education; the latter is a faith belonging to the constitution of man. Christ does not say that if we believe in His personal power to accomplish wonderful things we would be enabled to throw mountains into the ocean; but He spoke of our own faith, meaning the divine power of God in man, that may act through ourselves as much as it acted through Christ, if we become like Him. This power comes from God and returns to Him;

and if one man cures another in the name of Christ, he cures him by the power of God, and by his own faith. That power becomes active in and through him by his faith, and not out of gratitude for his professed belief, or the belief of the patient that Christ once existed upon the earth.

The power of the true faith extends as far as the power of God. Man can accomplish nothing by his own power of faith. If we did not have faith in our ability to walk, we would not be able to walk. If we accomplish anything whatever, faith accomplishes it through us.

Faith does not come from man, and no man can create faith; but faith is a power coming from God. Its germ is laid within man, and may be cultivated or neglected by him; it may be used by him for good or for evil, but it only acts effectively when it is strong and pure—not weakened by doubt, and not dispersed by secondary considerations. He who wants to employ it must have only *one* object in view. Diseases may be caused and cured by faith, and if men knew the power of faith they would have more faith and less superstition. We have no right to call a disease incurable; we have only the right to say that we cannot cure it. A physician who trusts only in his own science will accomplish little, but he who has faith in the power of God acting through him, and who employs that power intelligently, will accomplish much.

If any one thinks that he can cure a disease or accomplish anything else, merely because he believes that he is able to accomplish it, he believes in a superstition; but if he believes that he can perform such a thing, because he is conscious of having the power to do so, he will then be able to accomplish it by the power of the true faith. Such a faith is knowledge and gives power. True faith is spiritual consciousness, but a belief based upon mere opinions and creeds is the product of ignorance, and is superstition.

The body which we receive from our parents, and which is built up from the nutriments it draws directly and indirectly from the earth, has no spiritual powers, for wisdom and virtue, faith, hope, and charity, do not grow from the earth. They are not the products of man's physical organization, but the attributes of another invisible and glorified body, whose germs are laid within man. The physical body changes and dies, the glorified body is eternal. This eternal man is the real man, and is not generated by his earthly parents. He does not draw nutriment from the earth, but from the eternal invisible source from which he originated. Nevertheless the two bodies are one, and man may be compared to a tree, drawing his nutriments from the earth, and from the surrounding air. The roots extend into the earth, and seek their nutriment in the dark, but the leaves receive their nutriment from the light. The temporal body is the house of the eternal, and we should therefore take care of it, because he who destroys the temporal body destroys the house of the eternal, and although the eternal man is invisible, he exists nevertheless, and will become visible in time, just as a child in its mother's womb is invisible before it is born, but after its birth it may be seen by all but those who are blind; and as everything returns after a while to the source from whence it came, so the body returns to the earth and the spirit to heaven or hell.

CHAPTER 26

MICHEL DE MONTAIGNE

MICHEL DE MONTAIGNE WAS born on February 28, 1533, at the ancestral chateau near Bordeaux. He studied under some of the noted humanists of the day and read widely of law. At twenty-one he became a councillor in the Bordeaux courts, and when he resigned the post—after a number of years—he retired to his tower room although he lived at the French court for a time and was a favorite with successive monarchs.

After 1571 he spent most of his time at his chateau, writing, publishing, and revising his Essays which were destined to influence the style and form of French prose in the subsequent era.

He wrote upon many subjects, and often drew his inspiration from the Ancients who had been brought so close to him through his schooling. But in one matter, at least, he was far ahead of his time. We are quoting him on that subject, through excerpts from his essay on education. It may be that these enlightened ideas were fostered by those of his father. At that time it was customary to whip students into an understanding of the "dead" languages and to eradicate all shortcomings by means of corporal punishment. Montaigne's father engaged a tutor who spoke no French, and instructed that no one converse with his son in any tongue but Latin. As a result he spoke it fluently before he was six, and had mastered it naturally without strain or punishment.

It is only within the last generation that civilized countries have begun to use the more advanced educational methods which Montaigne advocates in his essay, *On the Education of Children.*

THE TUTOR

'Tis the custom of schoolmasters to be eternally thundering in their pupils' ears, as they were pouring into a funnel, whilst the business of these is only to repeat what the others have said before. Now I would have a tutor to correct this error; and that, at the very first outset, he should, according to the capacity he has to deal with, put it to the test, permitting his pupil himself to taste and relish things, and of himself to choose and discern them, sometimes opening the way to him, and sometimes making him break the ice himself; that is, I would not have him alone to invent and speak, but that he should also hear his pupil speak in turn. Socrates, and, since him, Arcesilaus, made first their scholars speak, and then spoke to them. . . .

Let the tutor make his pupil examine and thoroughly sift every thing he reads, and lodge nothing in his head upon simple authority and upon trust. Let Aristotle's Principles be no more principles to him than those of Epicurus and the stoics; let the diversity of opinions be propounded to, and laid before him, he will himself choose, if he be able; if not, he will remain in doubt. . . .

For if he embrace the opinions of Xenophon and Plato, by the exercise of his reason they will no more be theirs, but become his own. Who follows another, follows nothing, finds nothing, nay, seeks nothing. . . . Let him, at least, know that he does know. 'Tis for him to imbibe their knowledge, but not to adopt their dogmas; and no matter if he forgets where he had his learning, provided he knows how to apply it to his own use; truth and reason are common to every one, and are no more his who spoke them first than his who spake them after. 'Tis no more according to Plato than according to me, since both he and I equally see and understand in the same manner. Bees cull their several sweets from this flower and that blossom, here and there where they find them, but themselves after make the honey which is all and purely their own, and no longer thyme and marjoram; so the several fragments the pupil borrows from others he will transform and blend together to compile a work that shall be absolutely his own; that is to say, his judgment, which his instruction, labour, and study should alone tend to form. He is not obliged to discover whence he had his materials, but only to produce what he has done with them.

. . .Who ever asked his pupil what he thought of grammar and rhetoric, or of such and such a sentence of Cicero. Our pedagogues stick them full feathered in our memories, and there establish them like oracles, of which the very letters and syllables are the substance of the thing. To know by rote is no knowledge, 'tis no more than only to retain what one has intrusted to his memory. That which a man rightly knows and understands he is the free disposer of at his own full liberty, without any regard to the author from whom he had it, or fumbling over the leaves of his book.

. . .But, withal, let my tutor remember to what end his instructions are principally directed, and that he do not so much imprint in his pupil's memory the date of the ruin of Carthage, as the manners of Hannibal and Scipio; nor so much where Marcellus died as why it was unworthy of his duty that he died there. Let him read history, not as an amusing narrative, but as a discipline of the judgment. 'Tis this study to which, in my opinion, of all others, we apply ourselves with the most differing and uncertain measures. I have read an hundred things in Livy, that another has not, or not taken notice of, at least; and Plutarch has read a hundred more than I could find, or than peradventure the author ever writ.

The Pupil

Let conscience and virtue be eminently manifest in his speech, and have only reason for their guide. Make him understand that to acknowledge the error he shall discover in his own argument, though only found out by himself, it is an effect of judgment and sincerity, which are the principal things he is to seek after. That obstinacy and contention are common qualities, most appearing in and best becoming a mean soul. That to recollect and correct himself, and to forsake a bad argument in the heights and heat of dispute, are great and rare philosophical qualities. Let him be directed, being in company, to have his eye and ear in every corner of the room; for I find that the places of greatest honour are commonly possessed by men that have least in them, and that the greatest fortunes are not always accompanied with the ablest parts. I have been present when, whilst they at the upper end of the table have been only commending the beauty of the arras, or the flavour of the wine, many fine things have been lost or thrown away at the lower end of the table. Let him examine every man's talent; a peasant, a bricklayer, or any casual passenger, a man may learn something from every one of these in their several capacities, and something will be picked out of their discourse, whereof some use may be made at one time or another; nay, even the folly and weakness of others will contribute to his instruction. By observing the graces and manners of all he sees, he will create to himself an emulation of the good, and a contempt of the bad. . . .

Since philosophy is that which instructs us to live, and that infancy has there its lessons as well as other ages, why is it not communicated to children betimes? . . . Philosophy has discourses equally proper for childhood as for old age. . . .

But to our young friend, a closet, a garden, the table, his bed, solitude, and company, morning and evening, all hours shall be the same, and all places to him a study; for philosophy, who as the formatrix of judgment and manners shall be his principal lesson, has that privilege to have a hand in everything. . . . By which method of instruction, my young pupil will be much more and better employed than those of the college are. But as the steps we take in walking to and fro in a gallery, though three times as many, do not tire a man so much as those we employ in a formal journey; so our lesson, concurring as it were accidentally, without any set obligation of time or place, and falling naturally in with every action, will insensibly insinuate itself. Our very exercises and recreations, running, wrestling, music, dancing, hunting, riding, and fencing, will prove to be a good part of our study. I would have his outward behaviour and mien, and the disposition of his limbs, formed at the same time with his mind. It is not a soul, it is not a body, that we are training up; it is a man, and we ought not to divide him into two parts; and, as Plato says, we are not to fashion one without the other, but make them draw together like two horses harnessed to a coach. . . .

As to the rest, this method of education ought to be carried on with a firm gentleness, quite contrary to the practice of our pedants, who instead of tempting and alluring children to letters, present nothing before them but rods and ferules, horror and cruelty. Away with this violence! away with this compulsion; than which, I certainly believe nothing more dulls and degenerates a well-born nature. If you would have him fear shame and chastisement, do not harden him to them. Inure him to heat and cold, to wind and sun, and to dangers that he ought to despise.

CHAPTER 27

SIR FRANCIS BACON

SIR FRANCIS BACON, ROSICRUCIAN Imperator, English philosopher, statesman, essayist, Lord Chancellor of England, and the recognized writer by many literary critics, of the Shakespearian plays. He was born at York House in the Strand, London, on January 22, 1561, of noble birth. In April 1573, he entered Trinity College, Cambridge. Fond of the sciences, he diligently applied himself to the various sciences of the time. Later he says of his real aspirations in the *de Interpretatione Naturae Prooemium*: "I found I was fitted for nothing so well as for the study of truth; as having a mind nimble and versatile enough to catch the resemblances of things and at the same time steady enough to seize and distinguish the subtler differences; as being gifted by nature with a desire to ask, patience to doubt, fondness to meditate, slowness to assert, readiness to consider, carefulness to dispose and set in order; and as being a man that neither effects what is new, nor admits what is old, and that hates every kind of imposter, so I thought my nature had a kind of familiarity and relation with truth."

In 1584 he took a seat in Parliament. At the close of 1591, Bacon was acting as the confidential adviser of the Earl of Essex, who was Elizabeth's favorite. Bacon proposed in 1613 to the King that Sir Edward Coke be advanced to the King's bench. This change of legal position opened the way for Bacon to become Attorney General. Four years later, in January of 1618, he became Lord Chancellor of England. His various literary works at that time were well read, and although severely criticized in some circles, were bringing him recognition. His most celebrated work was *Novum Organum.*

It is only in recent years that the true story of Bacon's accusation and trial has become known. Bacon's political power and prominence, as well as his literary ability, aroused considerable enmity and his enemies accused him of bribery. At first the charge was dropped. Later, the charges became so complicated, so extremely difficult to prove innocence, that he was tried and convicted. He was fined an enormous amount and sentenced to a long period in prison. However, the fine was removed and he only served four days of the sentence. He was permitted to again practice at the bar, although he never again held a seat in Parliament. He continued his literary work.

Researchers and biographers of the present say, as it is known also to Rosicrucians, "On the whole, it appears that Bacon's own account of this painful episode is substantially correct." He stated he had received offers of bribes which he had never accepted. He affirms that his intention was never swerved by a bribe, and in several cases, his judgment seems to have been rendered against the parties attempting to bestow the bribe. The remainder of his life was spent in a work far more valuable to the world than any thing he had accomplished in his earlier years, politically or legally. The following selection is from his *Novum Organum*, (New Instrument or Implement).

ENCOURAGING THE PURSUIT OF KNOWLEDGE

It seems to me that men do not rightly understand either their store or their strength, but overrate the one and underrate the other. Hence it follows, that either from an extravagant estimate of the value of the arts which they possess, they seek no further; or else from too mean an estimate of their own powers, they spend their strength in small matters and never put it fairly to the trial in those which go to the main. These are as the pillars of fate set in the path of knowledge; for men have neither desire nor hope to encourage them to penetrate further. And since opinion of store is one of the chief causes of want, and satisfaction with the present induces neglect of provision for the future, it becomes a thing not only useful, but absolutely necessary that the excess of honor and admiration with which our existing stock of inventions is regarded be in the very entrance and threshold of the work, and that frankly and without circumlocution, stripped off, and men be duly warned not to exaggerate or make too much of them. For let a man look carefully into all that variety of books with which the arts and sciences abound, he will find everywhere endless repetitions of the same thing, varying in the method of treatment, but not new in substance, insomuch that the whole stock, numerous as it appears at first view, proves on examination to be but scanty. And for its value and utility it must be plainly avowed that that wisdom which we have derived principally from the Greeks is but like the boyhood of knowledge, and has the characteristic property of boys; it can talk, but it cannot generate; for it is fruitful of controversies but barren of works. So that the state of learning as it now is, appears to be represented to the life in the old fable of Scylla, who had the head and face of a virgin, but her womb was hung round with barking monsters, from which she could not be delivered. For in like manner the sciences to which we are accustomed have certain general positions which are specious and flattering; but as soon as they come to particulars, which are as the parts of generation, when they should produce fruit and works, then arise contentions and barking disputations, which are the end of the matter and all the issue they can yield. Observe also, that if sciences of this kind had any life in them, that could never have come to pass which has been the case now for many ages—that they stand almost at a stay, without receiving any augmentations worthy of the human race, insomuch that many times not only what was asserted once is asserted still, but what was a question once is a question still, and instead of being resolved by discussion is only fixed and fed; and all the tradition and succession of schools is still a succession of masters and scholars, not of inventors and those who bring to further perfection the things invented. In the mechanical arts we do not find it so; they, on the contrary, as having in them some breath of life, are continually growing and becoming more perfect. As originally invented they are commonly rude, clumsy, and shapeless; afterwards they acquire new powers and more commodious arrangements and constructions; in so far that men shall sooner leave the study and pursuit of them and turn to something else, than they arrive at the ultimate perfection of which they are capable. Philosophy and the intellectual sciences on the contrary, stand like statues, worshipped and celebrated, but not moved or advanced. Nay, they sometimes flourish most in the hands of the first author, and afterwards degenerate. For when men have once made over their judgments to others' keeping, and (like those senators whom they called *Pedarii*) have agreed to support some one person's opinion, from that time they make no enlargement of the sciences themselves, but fall to the servile office of embellishing certain individual authors and increasing their retinue. And let it not be said that the sciences have been growing gradually till they have at last reached their full stature, and so (their course being completed) have settled in the works of a few writers; and that there being now no room for the invention of better, all that remains is to embellish and cultivate those things which have been invented already. Would it were so! But the truth is that this appropriating of the sciences has its origin in nothing better than the confidence of a few persons and the sloth and indolence of the rest. For after the sciences had been in several parts perhaps cultivated and handled diligently, there has risen up some man of bold disposition, and famous for methods and short ways which people like, who has in appearance reduced them to an art, while he has in fact only spoiled all that the others had done. And yet this is what posterity like, because it makes the work short and easy, and saves further inquiry, of which they are weary and impatient. And if any one take this general acquiescence and consent for an argument of weight, as being the judgment of Time, let me tell him that the reasoning on which he relies is most fallacious and weak. For, first, we are far from knowing all that in the matter of sciences and arts has in various ages and places been brought to light and published; much less, all that has been by private persons secretly attempted and stirred; so neither the births nor the miscarriages of Time are entered in our records. Nor, secondly, is the consent itself and the time it has continued a consideration of much worth. For however various are the forms of civil polities, there is but one form of polity in the sciences; and that always has been and always will be popular. Now the doctrines which find most favor with the populace are those which are either contentious and pugnacious, or specious and empty; such, I say, as either entangle assent or tickle it. And, therefore, no doubt the greatest wits in each successive age have been forced out of their own course; men of capacity and intellect above the vulgar having been fain, for reputation's sake, to bow to the judgment of the time and the multitude; and thus if any contemplations of a higher order took light anywhere, they were presently blown out by the winds of vulgar opinions. So that Time is like a river, which has brought down to us things light and puffed up, while those which are weighty and solid have sunk. Nay, those very authors who have usurped a kind of dictatorship in the sciences and taken upon them to lay down the law with such confidence, yet when from time to time they come to themselves again, they fall to complaints of the subtlety of nature, the hiding-places of truth, the obscurity of things, the entanglement of causes, the weakness of the human mind; wherein nevertheless they show themselves never the more modest, seeing that they will rather lay the blame upon the common condition of men and nature than upon themselves. And then whatever any art fails to attain, they ever set it down upon the authority of that art itself as impossible of attainment; and how can art be found guilty when it is judge in its own cause? So it is but a device for exempting ignorance from ignominy.

CHAPTER 28

JOHANNES KEPLER

JOHANNES KEPLER WAS BORN in the German "free city" of Weil der Stadt in 1571. He inherited very little from his father. The elder Kepler, just before his death, lost the bulk of his estate through endorsing papers for a friend. Kepler was sent to a monastic school and finally to the University of Tubingen. Physically, he was sickly; mentally, he was alert and ranked second in his class. He became interested in the Copernican system and through this interest became the assistant in 1599 of Tycho Brahe.

Kepler's great contributions to astronomy are due to this early association with Brahe, a Danish nobleman, a maker of astronomical instruments and the compiler of the famous *Rudolphine Tables*. Kepler's first assignment was the study of the planet Mars, and he continued this investigation after the death of Brahe. Kepler, however, was more than a scientist; he was a metaphysician. He was not satisfied alone with the observation of the movements of the Cosmic bodies, but he tried to formulate a metaphysical theory for their origin and their relationship.

Following are excerpts of Kepler's writings on the principles of astronomy. There is a masterful beauty to his writings. In their simplicity and forcefulness, you are conscious of a lack of technical phraseology and terms, yet it seems that you are being made acquainted with the profound mysteries of the universe.

ON THE PRINCIPLES OF ASTRONOMY

What is astronomy? It is the science of treating the causes of those celestial appearances which we who live on the earth observe and which mark the changes of times and seasons; by the studying of which we are able to predict for the future the face of the heavens, that is, the stellar phenomena, and to assign fixed dates for those which have occurred in the past.

Why is it called astronomy? From the law (nomos) or governance of the stars (astra), that is, of the motions in which the stars move, just as economy is named from the law of domestic affairs (oiconomia) and paedonomy (paidonomia) from the ruling of youths.

What is the relation of this science to the other sciences?

1. It is a branch of physics because it investigates the causes of natural objects and events, and because among its subjects are the motions of the heavenly bodies, and because it has the same end as physics, to inquire into the conformation of the world and its parts.

2. Astronomy is the soul of geography and hydrography, for the various appearances of the sky in various districts and regions of the earth and sea are known only by astronomy.

3. Chronology is dependent upon it, because the movements of the heavenly bodies prescibe seasons and years and date the histories.

4. Meteorology is also its subordinate, for the stars move and influence this sublunary nature and even men themselves.

5. It includes a large part of optics, because it has a subject in common with that; that is, the light of the heavenly bodies, and because it corrects many errors of sight in regard to the character of the earth and its motions.

6. It is, however, subordinate to the general subject of mathematics and uses arithmetic and geometry as its two wings, studying the extent and form of the bodies and motions of the universe and computing the periods, by these means expediting its demonstrations and reducing them to use and practical value.

How many, then, are the branches of astronomical study? The departments of the study of astronomy are five; historical, in the matter of observations, optical as to the hypothesis, physical as to the causes of the hypotheses, arithmetical as to the tables and calculations, mechanical as to its instruments.

* * *

Since we must begin with appearances, explain how the world seems to be made up? The world is commonly thought, accepting the testimony of the eyes, to be an immense structure consisting of two parts, the earth and the sky.

What do men imagine concerning the figure of the earth? The earth seems to be a broad plane extending in a circle in every direction around the spectator. And from this appearance of a plane bounded by a great circle the appellation, orbis terrarum, the circle of the earth, has arisen, and has been taken over by the Scripture and among other nations.

What do men imagine to be the center of the earth? Each nation, unless it has become familiar with the notion of the circle, thinks by the instinct of nature and the error of vision that its country is in the center or middle of this plane circle. So the common people among the Jews believe still that Jerusalem, the earliest home of their race, is situated at the center of the world.

What do men think about the waters? Since men proceeding as far as possible in any direction finally came upon the ocean, some have thought that the earth is like a disc swimming in the waters, and that the waters are held up by the lower part of the sky, whence poets have called the ocean, the father of all things. Others believe that a strip of land surmounts the ocean which keeps the water from flowing away, and these suppose there is land under the water, saying that the water is held up by the earth. Besides these there are still others who, since the ocean seems higher than the land if it is looked at from the edge of the shore, believe that the earth is, as it were, sunk in the water and supernaturally guarded by the omnipotence of God lest the waters rushing in from the deep should overwhelm it.

What do men imagine to be under both the land and the waters? There has been great discussion among men marveling concerning the foundation which could bear up the great mass of the earth so that it should remain for so many centuries firm and immovable and should not sink; and Heraclitus among the early philosophers, and Lactantius among the ecclesiastics said that it reached down to the lowest root of things.

How about the other part of the world, the sky and its extent? Men have thought that the sky was not much larger than the earth, and indeed was connected with the earth and the ocean at the circumference of the circle, so that it bounded the earth; and that anyone going that far, if it could be done, would run up against the sky, blocking further progress. With this idea of men the Scriptures also agreed.

So also the poets said that Mt. Atlas, a lofty mountain on the farthest shore of Africa, bore up the sky on his shoulders, and Homer placed the Aethiopeans at the extremities of the rising and setting sun, thinking that because of the contiguity of the earth and sky there, the sun was so close to them that it burned their skin.

What form do they ascribe to the sky? The eyes ascribe to the sky the shape of a tent, extending over our heads and beyond the sun, moon and stars, or rather the shape of an arch overspanning the terrestrial plane, with a long curve, so that the part of the sky just over the head of the spectator is much nearer to him than the part that touches the mountains.

What have men conceived in regard to the motion of the sky? Whether the sky moves or stands still is not apparent to the sight because the tenuity of its substance escapes the eyes, unless indeed those things appear to stand still in which the eye can perceive no variation. But the changing positions of the sun, moon and stars in relation to the ends of the earth was apparent to the eyes. For the sun seems to emerge from an opening between the sky and the immovable mountains and ocean, as if coming out of a chamber, and having traversed the vault of the sky seems to sink again in the opposite region; so also the moon, and the planets, and the whole host of stars proceed as if strictly marshaled and drawn up in line, first one and then the other marching alone, each in his order and place.

And so, since the ocean lies beyond the extreme lands, the mass of men have thought that the sun plunges into the ocean and is extinguished, and from the opposite region a new sun issues forth daily from the ocean. The poets have used this figure in their creations. But, indeed, there have been even philosophers who have declared that on the farthest shores of Lusitania could be heard the roar of the ocean extinguishing the flames of the sun, as Strabo recounts.

CHAPTER 29

JACOB BOEHME

JACOB BOEHME, WAS BORN in 1575 in the little town of Alt Seidenberg, then in Bohemia, now in East Germany. He was the son of poor, country folk, who were uneducated. He eventually went to school, merely to learn to read and write. After receiving that much education, he became an apprentice at a local shoemaker's shop. In 1599, by diligent application, he had become a master shoemaker. He married a local woman, and they had four sons.

In the year 1600, when he was 25 years of age, he had a Divine "Illumination." These periods of Illumination, even in his early youth, were quite frequent. At this Illumination, it is said he learned the innermost foundation of nature. Ten years later in 1610, he was again lifted to the heights in another Illumination. At this time, he recognized the Divine order of nature.

Between the years 1612 and 1624, he wrote many books about the things which he saw in the light of his own spirit. He experienced transition on Saturday, November 20, 1624. His view of transition was one that would be expected from a mystic, having no fear of death, and looking upon it as a greater "Initiation."

Jacob Boehme had many enemies because of his philosophical views, which were contrary to the bigoted, narrow, religious concepts of the day. Outstanding was the bigoted headparson, Gregorious Richter, who refused a decent burial to the corpse of the philosopher. Eventually, Count Hannibal von Drohna arrived in the community and ordered the body to be buried in a solemn manner. So prejudiced was the clergy, that the clergyman who gave the sermon began his speech by expressing his great disgust at having been forced to do so.

We find, therefore, in the instance of Jacob Boehme, another example of the attempt to belittle the expounder of profound knowledge and to suppress knowledge itself. Henry Moore, professor at Cambridge University, after examining the books of Jacob Boehme, said: "He who treated Boehme with contempt could not be otherwise but ignorant and mentally blind."

THREE PRINCIPLES

Man is a product of three worlds. His spirit is of God, his soul from the constellation of the astral elements, his body from the elements of the terrestrial plane. In each of these aspects he partakes of the attributes of the principle from which he has originated. As a spirit he is, and has been, and will always be, immortal; and is even now in heaven, from which he has never departed. As a product of the astral plane, he is subject to the conditions existing therein, while his physical form must dissolve again into the elements to which it belongs. With whatever of these three states man identifies himself, that state will be his own.

God willed to become manifest in all three principles, but the order did not remain as it was originally instituted. The middle went into the exterior, and the exterior into the middle. This is not the order of eternity, and, therefore, the external and the inner principles must become separated.

The life which we receive in the body of our mother is merely from the power of the sun, the stars, and the elements, which not only organize the body of the child and endow it with life, but which also bring it to light and nourish and nurse it during the whole term of its life. They likewise distribute to it fortune and misfortune, and finally they cause it to die and to decompose.

Behold what you are. Dust of the earth; a corpse. Your life is subject to the stars and elements. It is they who rule you according to their qualities, and they endow you with talents and arts; but when their period and constellation under which you have been conceived and born is ended, then they will forsake you.

The corporeal essences return to the earth; the elemental spirit, the air, returns to the air; the water and blood are received by the terrestrial water and earth, and there remains nothing of the external man. He has then ceased to exist. He had a beginning and he had an end.

At death the four elements separate from the one element. Then the tincture, together with the shadow of that which constituted the man, goes into the ether and remains within the root of that element from which the four elements were born, and from which they emanated.

After the death of the physical form man remains still a being of twofold aspect; namely, as a celestial spirit, according to the divine principle in him (of which he may or may not be conscious); and, secondly, as a supersensual, but, nevertheless, material being, according to his astral body. Each of these essences now gravitates to the plane to which it belongs according to its qualities. From this double but opposite tendency results the rupture or division of the soul and the judgment.

When a person in this world dies, he then comes before the angel who in his sword carries death and life, the love and wrath of God. There his soul has to pass through the judgment at the portals of Paradise. If she has been captured by the wrath of God she will not be able to pass through his door, but if she is a child of the virgin and born of the seed of the (celestial) woman, she will then pass through. Then will the angel cut away from her nature that which has been generated by the serpent, and the soul will then serve God in His holy temple in Paradise, waiting there for the resurrection of her (celestial) body.

During his terrestrial existence man can remain consciously in the three worlds, and by the power of the will with which he is endowed penetrate into either one or the other; but after the separation of the soul from the body has taken place, he can continue to exist as an individuality only in one of these worlds, either within the realm of divine light or within the power of the fire; because together with his physical body he loses the power of self-government. He can then no longer follow his own will but has to go where he gravitates.

There are three principles in the constitution of man, either of which he may unfold during his terrestrial existence; but after the body is disorganized, he then lives only in one principle and cannot evolve the other. In eternity he must remain in that state of consciousness which he had acquired here.

There are not three separate souls in man, but only one. This soul stands in three principles; namely, the realm of the wrath, the realm of the love of God and the kingdom of this world. When the air of the external kingdom of this world deserts the soul, then will she become manifest in either the dark realm of fire or in the holy kingdom of light, which is the kingdom of the love-fire, the power of God. To whatever plane she has surrendered herself during her earthly existence, therein she remains after the external kingdom has departed from her.

During his terrestrial life man may live either in heaven or hell, or come out of one of these states and enter the other, because he can then govern his will by means of his intellect; but after the death of the body the function of the brain necessary for that purpose exists no longer, and then the soul is not able to change her will. She therefore becomes absorbed entirely into that principle which has obtained ruling power within her own nature. For this reason, it is of paramount importance for man to seek to unfold during his terrestrial life the love of God; i.e., the appreciation of the ideal and the will to realize all that is noble and good within his soul, so that it will act as his guiding star in eternity.

Man is in this world already in heaven or hell, wherever he corporeally may be. If his spirit is in harmony with God, he is then spiritually in heaven and his soul is in God. If he spiritually dwells in the wrath, he is then already in hell and in company of all the devils.

Here in the life of the soul is the balance. If she is evil, she can be reborn in love; but when the balance breaks and the angel has turned, then will she be in that principle which is prevailing in her.

During her terrestrial life the soul can change her will, but after the death of the body there remains nothing within her power by which she can change her will.

Whatever the soul during her terrestrial life receives within her will, and wherewith she becomes entangled, that she will take in her will with her after the death of the body, and she can no more rid herself of it, because she has then nothing but that wherein she has entered, and which now constitutes her very self. But during terrestrial life she may destroy that wherein she has become entangled in her will.

CHAPTER 30

THOMAS HOBBES

THOMAS HOBBES WAS A materialist. He can hardly be thought of as a metaphysician; yet, his profound inquiry into the nature of the functions of man warrants our most careful consideration of his views. His reasoning is superb. As Rosicrucians we must admit that much of man is dependent upon the physical laws of the universe, and through these laws Hobbes has made man's material nature more thoroughly understood, even though we disagree in part with some of his doctrines.

Hobbes was born at Westport, Wiltshire, England, April 5, 1588. He was raised by his uncle; his father fled due to difficulties arising from a quarrel with a rival. He studied Greek and Latin and at the early age of fifteen entered Oxford, graduating in 1608. At that time he was just casually interested in scholastic learning. He became tutor and secretary to young Cavendish, Earl of Devonshire, until the death of Cavendish. Then he tutored his son. He formed the conception that everything in the universe—all of its forces, all of its manifestation—is the result of a universal motion, a physical energy and that mind in man was no exception.

In 1647 he became instructor to the Prince of Wales, during which time he published his great work, *Leviathan*. In his *Leviathan* he tried to apply the same principles to society as he applied to the creations of nature. "Society," he said, "was an organism, the basis of which was a contract between the people and the king—each accordingly benefitted." This philosophy of tracing everything that exists in the universe or that is known to man to purely mechanical principles devoid of intelligence, naturally offended the clergy, though he was restored to good favor after the Restoration.

Thomas Hobbes was one of the new school of materialistic philosophers, who went a long way toward contributing to our present advance in science. The pendulum is swinging the other way today. The origin of everything that is cannot alone be explained by physical laws. It is necessary to take into consideration a primary cause, which cause day by day is more compatible with the metaphysical principle of infinite intelligence. Below you will find excerpts from his writings in the *Leviathan*.

OF IMAGINATION

That when a thing lies still, unless somewhat else stir it, it will lie still for ever, is a truth that no man doubts of. But that when a thing is in motion, it will eternally be in motion, unless somewhat else stay it, though the reason be the same, namely, that nothing can change itself, is not so easily assented to. For men measure, not only other men, but all other things, by themselves; and because they find themselves subject after motion to pain, and lassitude, think every thing else grows weary of motion, and seeks repose of its own accord; little considering, whether it be not some other motion, wherein that desire of rest they find in themselves, consisteth. From hence it is, that the schools say, heavy bodies fall downwards, out of an appetite to rest, and to conserve their nature in that place which is most proper for them; ascribing appetite, and knowledge of what is good for their conservation, which is more than man has, to things inanimate, absurdly.

When a body is once in motion, it moveth, unless something else hinder it, eternally; and whatsoever hindreth it, cannot in an instant, but in time, and by degrees, quite extinguish it; and as we see in the water, though the wind cease, the waves give not over rolling for a long time after; so also it happeneth in that motion, which is made in the internal parts of a man, then, when he sees, dreams, etc. For after the object is removed, or the eye shut, we still retain an image of the thing seen, though more obscure than when we see it. And this is it, the Latins call *imagination*, from the image made in seeing; and apply the same, though improperly, to all the other senses. But the Greeks call it *fancy*; which signifies *appearance*, and is as proper to one sense, as to another. *Imagination* therefore is nothing but *decaying sense*; and is found in men, and many other living creatures, as well sleeping, as waking.

The decay of sense in men waking, is not the decay of the motion made in sense; but an obscuring of it, in such manner as the light of the sun obscureth the light of the stars; which stars do no less exercise their virtue, by which they are visible, in the day than in the night. But because amongst many strokes, which our eyes, ears, and other organs receive from external bodies, the predominant only is sensible; therefore, the light of the sun being predominant, we are not affected with the action of the stars. And any object being removed from our eyes, though the impression it made in us remain, yet other objects more present succeeding, and working on us, the imagination of the past is obscured, and made weak, as the voice of a man is in the noise of the day. From whence it followeth, that the longer the time is, after the sight or sense of any object, the weaker is the imagination. For the continual change of man's body destroys in time the parts which in sense were moved; so that distance of time, and of place, hath one and the same effect in us. For as at a great distance of place, that which we look at appears dim, and without distinction of the smaller parts; and as voices grow weak, and inarticulate; so also, after great distance of time, our imagination of the past is weak; and we lose, for example, of cities we have seen, many particular streets, and of actions, many particular circumstances. This *decaying sense,* when we would express the thing itself, I mean *fancy* itself, we call *imagination*, as I said before: but when we would express the decay, and signify that the sense is fading, old, and past, it is called *memory*. So that imagination and memory are but one thing, which for divers considerations hath divers names.

Much memory, or memory of many things, is called *experience*. Again, imagination being only of those things which have been formerly perceived by sense, either all at once, or by parts at several times; the former, which is the imagining the whole object as it was presented to the sense, is *simple* imagination, as when one imagineth a man, or horse, which he hath seen before. The other is *compounded* as when, from the sight of a man at one time, and of a horse at another, we conceive in our mind a Centaur. So when a man compoundeth the image of his own person with the image of the actions of another man, as when a man imagines himself a Hercules or an Alexander, which happeneth often to them that are much taken with reading of romances, it is a compound imagination, and properly but a fiction of the mind. There be also other imaginations that rise in men, though waking, from the great impression made in sense; as from gazing upon the sun, the impression leaves an image of the sun before our eyes a long time after; and from being long and vehemently attent upon geometrical figures, a man shall in the dark, though awake, have the images of lines and angles before his eyes; which kind of fancy hath no particular name, as being a thing that doth not commonly fall into men's discourse.

The imaginations of them that sleep are those we call *dreams*. And these also, as all other imaginations, have been before, either totally or by parcels, in the sense. And because in sense, the brain and nerves, which are the necessary organs of sense, are so benumbed in sleep, as not easily to be moved by the action of external objects, there can happen in sleep no imagination, and therefore no dream, but what proceeds from the agitation of the inward parts of man's body; which inward parts, for the connection they have with the brain, and other organs, when they be distempered, do keep the same in motion; whereby the imaginations there formerly made, appear as if a man were waking; saving that the organs of sense being now benumbed, so as there is no new object, which can master and obscure them with a more vigorous impression, a dream must needs be more clear, in this silence of sense, than our waking thoughts. And hence it cometh to pass, that it is a hard matter, and by many thought impossible, to distinguish exactly between sense and dreaming. For my part, when I consider that in dreams I do not often nor constantly think of the same persons, places, objects, and actions, that I do waking; nor remember so long a train of coherent thoughts, dreaming, as at other times; and because waking I often observe the absurdity of dreams, but never dream of the absurdities of my waking thoughts; I am well satisfied, that being awake, I know I dream not, though when I dream I think myself awake.

CHAPTER 31

JOHANN AMOS COMENIUS

JAN AMOS KOMENSKY, IN Latinized form, Johann Amos Comenius, was born in 1592 in Moravia (modern Czechoslovakia). He was orphaned early in life. He completed elementary school at Stassnick, and at the age of sixteen entered the Latin school. His higher education was obtained at the universities of Amsterdam and Heidelberg. He was appointed in 1618 head of the Moravian school and church at Fulneck. In 1627, after losing his property due to Spanish troops, he sought peace and seclusion in Poland. In 1641 he was summoned to England to add his experience to the improvement of the public schools, but civil war prevented him from completing his work. He returned to Poland and in a few years founded a model school. In 1670 he died at the home of Lawrence de Geer.

He patterned his method of education after nature. He made a close study of the manner in which nature progresses and matures, and how different living things react to their environment and he endeavored to adopt these principles to his public school system and to his plan for a model school.

Rosicrucians and all students will find his analysis of the method of acquiring knowledge and the way of preparing for study most helpful. Those who find it difficult to study, or who wish to know the best plan for study should especially give attention to this wonderful article of several centuries ago by Comenius.

THE PRINCIPLES OF FACILITY IN TEACHING AND IN LEARNING

We have already considered the means by which the educationist may attain his goal with certainty, we will now proceed to see how these means can be suited to the minds of the pupils, so that their use may be easy and pleasant.

Following in the footsteps of nature we find that the process of education will be easy

a. If it begin early, before the mind is corrupted.

b. If the mind be duly prepared to receive it.

c. If it proceed from the general to the particular.

d. And from what is easy to what is more difficult.

e. If the pupil be not overburdened by too many subjects.

f. And if progress be slow in every case.

g. If the intellect be forced to nothing to which the natural bent does not incline it, in accordance with its age and with the right method.

h. If everything be taught through the medium of the senses.

i. And if the use of everything taught be continually kept in view.

j. If every thing be taught according to one and the same method.

These, I say, are the principles to be adopted if education is to be easy and pleasant.

First Principle

Nature begins by a careful selection of materials. For instance, for hatching a bird she selects fresh eggs and those that contain pure matter. If the formation of the chicken have already begun, it is in vain to expect any result.

Imitation.—The architect who wishes to erect a building, needs a clear plot of ground, and, if there be a house already standing there, he must pull it down before he can build the new one.

The artist, too, does his best work on a clean canvas. If it have already been painted on, or be dirty or rough, it must be cleaned or smoothed before he can use it.

For the preservation of precious ointments, empty jars must be procured, or those that are in use must be carefully cleansed of their contents.

The gardener, too, prefers to plant young trees, or, if he takes them too old, cuts off the branches in order that the sap may not be dissipated. For this reason Aristotle placed "privation" among the principles of nature, for he held that it was impossible to impress a new form on any material until the old one had been removed.

Deviation.—It follows from this: (1) That it is best to devote the mind to the pursuit of wisdom while it is still fresh, and before it has acquired the habit of dissipating its strength over a variety of occupations; and that the later the education begins, the harder it will be for it to obtain a hold, because the mind is already occupied by other things. (2) That the result must be bad if a boy be instructed by several teachers at once, since it is scarcely possible for them all to use the same method, and, if they do not, the boy's mind is drawn first in one direction and then in another, and its development is thus hindered. (3) That it shows great lack of judgment if moral instruction be not made the first point when the education of children or of older boys is commenced; since, when they have been taught to control their feelings, they will be the more fit to receive other instruction. Horse-tamers keep a horse under absolute control with an iron bit, and ensure its obedience before they teach it its paces. Rightly does Seneca say: "First learn virtue, and then wisdom, since without virtue it is difficult to learn wisdom." And Cicero says: "Moral philosophy makes the mind fit to receive the seeds of further knowledge."

Rectification.—Therefore

a. Education should be commenced early.

b. The pupil should not have more than one teacher in each subject.

c. Before anything else is done, the morals should be rendered harmonious by the master's influence.

Second Principle

Nature prepares its material so that it actually strives to attain the form.

Thus the chicken in the egg, when sufficiently formed, seeks to develop itself still further, moves, and bursts the shell or breaks through it with its beak. After escaping from its prison, it takes pleasure in the warmth and nutriment provided by its mother, opens its beak expectantly and swallows its food greedily. It rejoices to find itself under the open sky, exercises its wings, and later on, uses them with enjoyment; in a word, it displays a keen desire to fulfill all its natural functions, though throughout the whole process of development it advances step by step.

Imitation.—The gardener also must bring it about that the plant, properly provided with moisture and with warmth, take pleasure in its vigorous growth.

Deviation.—Therefore, those who drive boys to their studies, do them great harm. For what result can they expect? If a man have no appetite, but yet takes food when urged to do so, the result can only be sickness and vomiting, or at least indigestion and indisposition. On the other hand, if a man be hungry, he is eager to take food, digests it readily, and easily converts it into flesh and blood. Thus Isocrates says: "He who is anxious to learn will also be learned."

And Quintilian says: "The acquisition of knowledge depends on the will to learn, and this cannot be forced."

Rectification.—Therefore

a. The desire to know and to learn should be excited in boys in every possible manner.

b. The method of instruction should lighten the drudgery of learning, that there may be nothing to hinder the scholars or deter them from making progress with their studies.

The desire to learn is kindled in boys by parents, by masters, by the school, by the subjects of instructions, by the methods of teaching, and by the authority of the state.

By parents, if they praise learning and the learned in the presence of their children, or if they encourage them to be industrious by promising them nice books and clothes, or some other pretty things; if they commend the teachers (especially him to whom they entrust their sons) as much for their friendly feeling towards the pupils as for their skill in teaching (for love and admiration are the feelings most calculated to stimulate a desire for imitation); finally, if, from time to time, they send the child to him with a small present. In this way they will easily bring it about that the children like their lessons and their teachers, and have confidence in them.

By the teachers, if they are gentle and persuasive, and do not alienate their pupils from them by roughness, but attract them by fatherly sentiments and words; if they commend the studies that they take in hand on account of their excellence, pleasantness, and ease; if they praise the industrious ones from time to time (to the little ones they may give apples, nuts, sugar, etc.); if they call the children to them, privately or in the class, and show them pictures of the things that they must learn, or explain to them optical or geometrical instruments, astronomical globes, and such like things that are calculated to excite their admiration; or again, if they occasionally give the children some message to carry to their parents. In a word, if they treat their pupils kindly they will easily win their affections, and will bring it about that they prefer going to school to remaining at home.

The school itself should be a pleasant place, and attractive to the eye both within and without. Within, the room should be bright and clean, and its walls should be ornamented by pictures. These should be either portraits of celebrated men, geographical maps, historical plans, or other ornaments. Without, there should be an open place to walk and to play in (for this is absolutely necessary for children, as we shall show later), and there should also be a garden attached, into which the scholars may be allowed to go from time to time and where they may feast their eyes on trees, flowers and plants. If this be done, boys will, in all probability, go to school with as much pleasure as to fairs, where they always hope to see and hear something new.

The subjects of instruction themselves prove attractive to the young, if they are suited to the age of the pupil and are clearly explained; especially if the explanation be relieved by a humorous or at any rate by a less serious tone. For thus the pleasant is combined with the useful.

If the method is to excite a taste for knowledge, it must, in the first place, be natural. For what is natural takes place without compulsion. Water need not be forced to run down a mountain-side. If the dam, or whatever else holds it back, be removed, it flows down at once. It is not necessary to persuade a bird to fly; it does so as soon as the cage is opened. The eye and the ear need no urging to enjoy a fine painting or a beautiful melody that is presented to them. In all these cases it is more often necessary to restrain than to urge on. The requisites of a natural method are evident from the preceding chapter and from the rules that follow.

In the second place, if the scholars are to be interested, care must be taken to make the method palatable, so that everything, no matter how serious, may be placed before them in a familiar and attractive manner; in the form of a dialogue, for instance, by pitting the boys against one another to answer and explain riddling questions, comparisons, and fables.

CHAPTER 32

RENÉ DESCARTES

RENÉ DESCARTES WAS AN eminent metaphysician and mathematician. He was born March 30, 1596, in the province of Touraine, France. Very shortly after his birth his mother died of consumption, and for a time it was believed that he too would not live. He was educated by the Jesuits at La Fleche from 1604 to 1612. Later he went to Paris to see life, life as it existed outside of the textbook. He spent two years there and later resumed his studies.

War eventually broke out and he enlisted as a volunteer against the Netherlands. While stationed in an army camp, having much time for reflection, he saw the possibilities of solving geometrical theorems by algebra, and from this concept was born analytical geometry. In 1641 he published his works known as *Meditationes de Prima Philosophia*. These tried to do away with all notions and start from one certain fact *Cogito, ergo sum* which means "I think and in thinking I exist." In other words, he tried to explain that so far as man is concerned, all reality, all there is to the world starts with man's consciousness, and to get a finer understanding of the universe, one must evolve his consciousness. He further concluded that matter's formation was purely a mechanical process. He tried to reduce the universe so far as its physical form is concerned to mechanics and the laws composing them. He was careful, however, not to permit this concept to spread too widely because of the religious opposition which would have arisen in the time in which he lived.

In 1649 he went to Stockholm on the invitation of Queen Christina, and while there caught cold and after a few weeks died on February 11, 1650. Below we give you excerpts from his *Meditations*. Compare his concept to Plato's doctrine of the ideas, Aristotle's doctrine of the ideas, and the philosophy of Locke.

MEDITATION I

OF THE THINGS WHICH WE MAY DOUBT

Several years have now elapsed since I first became aware that I had accepted even from my youth, many false opinions for true, and that consequently what I afterwards based on such principles was highly doubtful; and from that time I was convinced of the necessity of undertaking once in my life to rid myself of all the opinions I had adopted, and of commencing anew the work of building from the foundation, if I desired to establish a firm and abiding superstructure in the sciences. But as this enterprise appeared to me to be one of great magnitude, I waited until I had attained an age so mature as to leave me no hope that at any state of life more advanced I should be better able to execute my design. On this account, I have delayed so long that I should henceforth consider I was doing wrong were I still to consume in deliberation any of the time that now remains for action. Today, then, since I have opportunely freed my mind from all cares, (and am happily disturbed by no passions), and since I am in the secure possessions of leisure in a peaceable retirement, I will at length apply myself earnestly and freely to the general overthrow of all my former opinions. But, to this end, it will not be necessary for me to show that the whole of these are false—a point, perhaps, which I shall never reach; but as even now my reason convinces me that I ought not the less carefully to withold belief from what is not entirely certain and indubitable, than from what is manifestly false, it will be sufficient to justify the rejection of the whole if I shall find in each some ground for doubt. Nor for this purpose will it be necessary even to deal with each belief individually, which would be truly an endless labour; but, as the removal from below the foundation necessarily involves the downfall of the whole edifice, I will at once approach the criticism of the principles on which all my former beliefs rested.

All that I have, up to this moment, accepted as possessed of the highest truth and certainty, I received either from or through the senses. I observed, however, that these sometimes misled us; and it is the part of prudence not to place absolute confidence in that by which we have even once been declared.

But it may be said, perhaps, that, although the senses occasionally mislead us respecting minute objects, and such as are so far removed from us as to be beyond the reach of close observation, there are yet many other informations (presentations), of the truth of which it is manifestly impossible to doubt; as for example, that I am in this place, seated by the fire, clothed in a winter dressing gown, that I hold in my hands this piece of paper, with other intimations of the same nature. But how could I deny that I possess these hands and this body, and withal escape being classed with persons in a state of insanity, whose brains are so disordered and clouded by dark bilious vapors as to cause them pertinaciously to assert that they are monarchs when they are in the greatest poverty; or clothed (in gold) and purple when destitute of any covering; or that their head is made of clay, their body of glass, or that they are gourds? I should certainly be not less insane than they were I to regulate my procedure according to examples so extravagant.

Though this be true, I must nevertheless here consider that I am a man, and that, consequently, I am in the habit of sleeping, and representing to myself in dreams those same things, or even sometimes others less probable, which the insane think are presented to them in their waking moments. How often have I dreamt that I was in these familiar circumstances,—that I was dressed, and occupied this place by the fire, when I was lying undressed in bed? At the present moment, however, I certainly look upon this paper with eyes wide awake; the head which I now move is not asleep; I extend this hand consciously and with express purpose, and I perceive it; the occurrences in sleep are not so distinct as all this. But I cannot forget that, at other times, I have been deceived in sleep by similar illusions; and, attentively considering those cases, I perceive so clearly that there exist no certain marks by which the state of waking can ever be distinguished from sleep, that I feel greatly astonished; and in amazement I almost persuade myself that I am now dreaming.

Let us suppose, then, that we are dreaming, and that all these particulars—namely, the opening of the eyes, the motion of the head, the forth-putting of the hands—are merely illusions; and even that we really possess neither an entire body nor hands such as we see. Nevertheless, it must be admitted at least that the objects which appear to us in sleep are, as it were, painted representations which could not have been formed unless in the likeness of realities; and, therefore, that those general objects, at all events,—namely, eyes, a head, hands, and an entire body—are not simply imaginary, but really existent. For, in truth, painters themselves, even when they study to represent sirens and satyrs by forms the most fantastic and extraordinary, cannot bestow upon them natures absolutely new, but can only make a certain medley of the members of different animals; or if they chance to imagine something so novel that nothing at all similar has ever been seen before, and such as is, therefore, purely fictitious and absolutely false, it is at least certain that the colours of which this is composed are real.

And on the same principle, although these general objects, viz; (a body) eyes, a head, hands, and the like, be imaginary, we are nevertheless absolutely necessitated to admit the reality at least of some other objects still more simple and universal than these, of which, just as of certain real colours, all those images of things, whether true or real, or false and fantastic, that are found in our consciousness (cogitatio), are formed.

To this class of objects seem to belong corporeal nature in general and in its extension; the figure of extended things, their quantity and magnitude, and their number, as also the place in, and the time during, which they exist, and other things of the same sort. We will not, therefore, perhaps reason illegitimately if we conclude from this that Physics, Astronomy, Medicine, and all other sciences that have for their end the consideration of composite objects, are indeed of a doubtful character; but that Arithmetic, Geometry, and the other sciences of the same class, which regard merely the simplest and most general objects, and scarcely inquire whether or not these are really existent, contain somewhat that is certain and indubitable: for whether I am awake or dreaming, it remains true that two and three make five, and that a square has but four sides; nor does it seem possible that truths so apparent can ever fall under a suspicion of falsity (or incertitude).

Nevertheless, the belief that there is a God who is all-powerful, and who created me, such as I am, has for a long time, obtained steady possession of my mind. How, then, do I know that He has not arranged that there should be neither earth, nor sky, nor any extended thing, nor figure, nor magnitude, nor place, providing at the same time, however, for (the rise in me of the perceptions of all these objects, and) the persuasion that these do not exist otherwise than as I perceive them? And further, as I sometimes think that others are in error respecting matters of which they believe themselves to possess a perfect knowledge, how do I know that I am not also deceived each time that I add together two and three, or number the sides of a square, or form some judgment still more simple, if more simple indeed can be imagined? But perhaps Deity has not been willing that I should be thus deceived, for He is said to be supremely good. If, however, it were repugnant to the goodness of Deity to have created me subject to constant deception, it would seem likewise to be contrary to His goodness to allow me to be occasionally deceived; and yet it is clear that this is permitted. Some, indeed, might perhaps be found who would be disposed rather to deny the existence of a Being so powerful than to believe that there is nothing certain. But let us for the present refrain from opposing this opinion, and grant that all which is here said of Deity is fabulous; nevertheless, in whatever way it be supposed that I reached the state in which I exist, whether by fate, or chance, or by an endless series of antecedents and consequents, or by any other means, it is clear (since to be deceived and to err is a certain defect) that the probability of my being so imperfect as to be the constant victim of deception, will be increased exactly in proportion as the power possessed by the cause, to which they assign my origin, is lessened. To these reasonings I have assuredly nothing to reply, but am constrained at last to avow that there is nothing of all that I formerly believed to be true of which it is impossible to doubt, and that not through thoughtlessness or levity, but from cogent and maturely considered reasons; so that henceforward, if I desire to discover anything certain, I ought not the less carefully to refrain from assenting to those same opinions than to what might be shown to be manifestly false.

But it is not sufficient to have made these observations; care must be taken likewise to keep them in remembrance. For those old and customary opinions perpetually recur—long and familiar usage giving them the right of occupying my mind, even almost against my will, and subduing my belief; nor will I lose the habit of deferring to them and confiding in them so long as I shall consider them to be what in truth they are, viz., opinions to some extent doubtful, as I have already shown, but still highly probable, and such as it is much more reasonable to believe than deny. It is for this reason I am persuaded that I shall not be doing wrong, if, taking an opposite judgment of deliberate design, I become my own deceiver, by supposing, for a time, that all those opinions are entirely false and imaginary, until at length, having thus balanced my old by my new prejudices, my judgment shall no longer be turned aside by perverted usage from the path that may conduct to the perception of truth.

For I am assured that, meanwhile, there will arise neither peril nor error from this course, and that I cannot for the present yield too much to distrust since the end I now seek is not action but knowledge.

I will suppose, then, not that Deity, who is sovereignly good and the fountain of truth, but that some malignant demon, who is at once exceedingly potent and deceitful, has employed all his artifices to deceive me; I will suppose that the sky, the air, the earth, colours, figures, sounds, and all external things, are nothing better than illusions of dreams, by means of which this being has laid snares for my credulity; I will consider myself as without hands, eyes, flesh, blood, or any of the senses, and as falsely believing that I am possessed of these; I will continue resolutely fixed in this belief, and if indeed by this means it be not in my power to arrive at the knowledge of truth, I shall at least do what is in my power, viz., (suspend my judgment) and guard with settled purpose against giving my assent to what is false, and being imposed upon by this deceiver, whatever be his power and artifice.

But this undertaking is arduous, and a certain indolence insensibly leads me to my ordinary course of life; and just as the captive, who, perchance, was enjoying in his dreams an imaginary liberty, when he begins to suspect that it is but a vision, dreams awakening, and conspires with the agreeable illusions that the deception may be prolonged, so I, of my own accord, fall back into the train of my former beliefs, and fear to arouse myself from my slumber, lest the time of laborious wakefulness that would succeed this quiet rest, in place of bringing any light of day, should prove inadequate to dispel the darkness that will arise from the difficulties that have now been raised.

CHAPTER 33

THOMAS VAUGHAN

THOMAS VAUGHAN WAS A REAL occult philosopher, and of special interest to our readers is the fact that he was a member of the Rosicrucian Brotherhood. Below you will find excerpts taken from one of his works, the first selection being entitled, *A Letter from the Brothers of the R. C.* Thomas Vaughan was a brother of the English poet, Henry Vaughan. He, himself, was a poet but is better known for his mystical writings than for his verse. He lived in the seventeenth century. Most of his works were under the *nom de plume* of Eugenius Philalethes. Although he was ordained in the Church of England, he is known to have actively participated in the Great Rebellion. He was an ardent Royalist, and naturally sided with the King against the Commons. He was widely read; that is, as wide reading went during his epoch, and is also known to have lead a vagrant life.

His most important work was entitled *Lumen de Lumine,* from which the following excerpts are taken. His book, considered quite rare, was worded in a veiled manner as was customary at the time. He used the peculiar artifice of language common to the occult and mystical writers of his period, caring not the least whether the mass or multitude understood him. In fact, his work *Lumen de Lumine* was intended for mystical and philosophical students only—those who were more or less well versed with the subject. As a Rosicrucian and mystic, he was an active alchemist. It was only with the later growing interest in metaphysics that his work became generally known and appreciated. His works during his time were far in advance of the ability of the multitude to appreciate. When reading the following excerpts, you must realize that the peculiar terminology is an attempt to veil in allegorical form many truths. The student, the thinker, the Rosicrucian of today, may appreciate the simplicity of the modern presentation of the teachings in contrast to the manner in which they were presented in the past. This is a very evident indication of our freedom of thought. Freedom develops simplicity. Secrecy develops complexity.

A LETTER FROM THE BROTHERS OF THE R. C.

Every man naturally desires a superiority, to have treasures of gold and silver, and to seem great in the eyes of the world. God indeed created all things for the use of man, that he might rule over them and acknowledge therein the singular goodness and omnipotence of God, give Him thanks for His benefits, honour Him and praise Him. But there is no man who looks after these things otherwise than by spending their days idly; they would enjoy them without any previous labour and danger, neither do they look them out of that place where God hath treasured them up, Who expects also that man should seek for them there, and to those that seek will He give them. But there is not any that labours for a possession in that place, and therefore these riches are not found. For the way to this place, and the place itself, hath been unknown for a long time, and it is hidden from the greatest part of the world. But, notwithstanding it be difficult and laborious to find out this way and place, yet the place should be sought after. But it is not the will of God to conceal anything from those that are His, and therefore in this last age, before the final judgment comes, all these things shall be manifested to those that are worthy. As He Himself, though obscurely, lest it should be manifested to the unworthy, hath spoken in a certain place: there is nothing covered that shall not be revealed, and hidden that shall not be known. We, therefore, being moved by the Spirit of God, do declare the will of God to the world, which we have also already performed, and published in several languages. But most men either revile or contemn that our manifesto, or else, waiving the spirit of God, they expect the proposals thereof from us, supposing we will straightway teach them how to make gold by art, or furnish them with ample treasures, whereby they may live pompously in the face of the world—swagger and make wars—turn usurers, gluttons and drunkards, live unchastely, and defile their whole life with several other sins, all which things are contrary to the blessed will of God. These men should have learnt from those ten virgins—whereof five that were foolish demanded oil for their lamps from those five that were wise—how that the case is much otherwise. It is expedient that every man should labour for this treasure by the assistance of God and his own particular search and industry. But the perverse intentions of these fellows we understand out of their own writings, by the singular grace and revelation of God; we do stop our ears and wrap ourselves, as it were, in clouds, to avoid the bellowings and howlings of those men who in vain cry out for gold. And hence indeed it comes to pass that they brand us with infinite calumnies and slanders, which, notwithstanding, we do not resent, but God in His good time will judge them for it. But after that we had well known—though unknown to you—and perceived also by your writing how diligent you are to peruse the Holy Scripture and seek the true knowledge of God, we have also—above many thousands—thought you worthy of some answer, and we signify this much to you by the will of God and the admonition of the Holy Ghost.

There is a Mountain situated in the midst of the earth, or centre of the world, which is both small and great. It is soft—also above measure hard and stony; it is far off and near at hand; but, by the providence of God, invisible. In it are hidden most ample treasures, which the world is not able to value. This Mountain, by envy of the devil, who always opposeth the glory of God and the happiness of man, is compassed about with very cruel beasts and ravenous birds, which make the way thither both difficult and dangerous; and therefore hitherto, because the time is not yet come, the way thither could not be sought after nor found out. But now at last the way is to be found by those that are worthy, but notwithstanding, by every man's self-labour and endeavours.

To this Mountain you shall go in a certain night—when it comes—most long and most dark, and see that you prepare yourselves by prayer. Insist upon the way that leads to the Mountain, but ask not of any man where the way lies; only follow your Guide, who will offer himself to you, and will meet you in the way; but you shall not know him. This Guide will bring you to the Mountain at midnight, when all things are silent and dark. It is necessary that you arm yourselves with a resolute, heroic courage, lest you fear those things that will happen, and so fall back. You need no sword, nor any other bodily weapons; only call upon God sincerely and heartily. When you have discovered the Mountain, the first miracle that will appear is this—a most vehement and very great wind that will shake the Mountain and shatter the rocks to pieces. You shall be encountered also by lions and dragons and other terrible beasts; but fear not any of these things. Be resolute, and take heed that you return not, for your guide, who brought you thither, will not suffer any evil to befall you. As for the Treasure, it is not yet discovered, but it is very near. After this wind will come an earthquake, that will overthrow those things which the wind hath left, and make all flat. But be sure that you fall not off. The earthquake being past, there shall follow a fire that will consume the earthly rubbish and discover the Treasure; but as yet you cannot see it. After all these things, and near the daybreak, there shall be a great calm, and you shall see the day-star arise and the dawning will appear, and you shall perceive a great Treasure. The chiefest thing in it, and the most perfect, is a certain exalted Tincture, with which the world—if it served God, and were worthy of such gifts—might be tinged and turned into most pure gold.

* * * * *

THE PRESTER OF ZOROASTER

It is a miracle to consider how the earth, which is a body of inexpressible weight and heaviness, can be supported in the air, a fleeting, yielding substance, and through which even froth and feathers will sink and make their way. I hope there is no man so mad as to think it is poised there by some geometrical knack, for that were artificial, but the work of God is vital and natural. Certainly if the animation of the world be denied, there must needs follow a precipitation of this element by its own corpulency and gravity. We see that our own bodies are supported by that essence by which they are actuated and animated, but when that essence leaves them they fall to the ground till the spirit returns at the resurrection. I conclude then that the earth hath in her a fire-soul, a most powerful, strong spirit, that bears her up, as the spirit of man bears up man. To this agrees Raymond Lulley in the seventy-sixth chapter of his *Theory*: "The whole earth," saith he, "is full of intelligence, inclined to the discipline or operation of Nature, which intelligence is moved by the Superior Nature, so that the inferior intelligence is like to the Superior." This spirit or intelligence is the Prester, a notion of the admirable Zoroaster, as I find him rendered by Julian the Chaldean. It comes from *uro*, I burn, and signifies lightning or a certain burning *Turba*, or whirlwind, but in the sense of our Chaldee it is the fire-spirit of life. It is an influence of the Almighty God, and it comes from the Land of the Living Ones, namely, the Second Person, Whom the Kabalists style the Supernatural East. For as the natural light of the sun is first manifested to us in the East, so the Supernatural Light was first manifested in the Second Person, for He is *Principium Alterationis*, the Beginning of the Ways of God, or the First Manifestation of His Father's Light in the Supernatural Generation. From this *Terra Viventium*, or Land of the Living, comes all Life or Spirit, according to that position of the *Mekkubalim*: "Every good soul is a new soul, coming from the East," that is from *Chokmah*, or the Second *Sephira*, which is the Son of God.

Now for the better understanding of this descent of the soul, we must refer ourselves to another *placet* of the Kabalists, and this is it: "The souls," say they, "descend from the Third Light to the fourth day, thence to the fifth, whence they pass out and enter the night of the body." To understand this maxim, you must know there are three Supreme Lights or *Sephiroth*, which the Kabalist calls "one throne wherein sits the Holy, Holy, Holy Lord God of Hosts." This Third Light, from whence the souls descend, is Binah, or understanding, the last of the three *Sephiroth*, and it signifies the Holy Ghost. Now, that you may know in what sense this descent proceeds from that Blessed Spirit, I will somewhat enlarge my discourse, for the Kabalists are very obscure on the point. "To breathe is the property of the Holy Ghost," say the Jews. Now, we read that God breathed into Adam the Breath of Life, and he became a living soul. Here you must understand that the Third Person is the last of the Three, not that there is any inequality in them, but it is so in order of operation, for He applies first to the creature, and therefore works last. The meaning of it is this: The Holy Ghost could not breathe a soul into Adam but He must either receive it or have it of Himself. Now, the truth is He received it, and what He receives, that He breathes into Nature. Hence this most Holy Spirit is styled by the Kabalists the River flowing forth from Paradise, because He breathes as a river streams. He is called also *Mater filiorum*, Mother of sons, because by this breathing He is, as it were, delivered of those souls which have been conceived ideally in the Second Person. Now, that the Holy Ghost receives all things from the Second Person is confirmed by Christ Himself: "When the Spirit of Truth is come, He will guide you into all truth, for He shall not speak of Himself, but whatsoever He shall hear, that shall He speak, and He will show you things to come. He shall glorify me, for He shall receive of mine, and shall show it unto you. All things that the Father hath are mine; therefore, said I that He shall take of mine." Here, we plainly see, there is a certain subsequent order or merit in the operations of the Blessed Trinity, for Christ tells us that he receives from His Father, and the Holy Ghost receives from Him. Again, that all things are conceived ideally—or, as we commonly express it, created—by the Second Person is confirmed by the word of God: "The world was made by Him," saith the Scripture, "and the world knew Him not. He came unto His own, and His own received Him not."

CHAPTER 34

BARUCH SPINOZA

BARUCH SPINOZA, ALSO KNOWN as Benedictus de Spinoza, was born of a Jewish family in Amsterdam, Holland, in 1632. His parents had fled from the religious persecution prevalent in Spain at that time to Holland. As was customary among Jews in Holland, he received his early schooling under the tutorship of a rabbi.

He continued his schooling under a Christian teacher and lived for a time in Rijnsburg among the Collegiants, a group that has been compared to both the Quakers and the Pietists.

He developed a keen interest in languages, especially Latin, and became a fair linguist. The sciences also appealed to him and he devoted a great deal of time to them, as well as to philosophy. His study of philosophy brought him into conflict with the Jewish community. He entered into many debates with the officials of the orthodox faith in an attempt to defend his liberal views, but it resulted in his expulsion from the synagogue.

Among students of philosophy, the study of Descartes' writings was popular. Descartes was a materialist; to him the entire physical world, the material world, was the mechanical result of the principle of motion; mind was of God, but it was entirely separated from the material world; there was distinct dualism. To Spinoza both mind and matter were of God and were interrelated. Though at all times maintaining their distinct characteristics; they were two parallel creations of God, complementing each other. Spinoza spent most of his life on his meditations and correspondence, supporting himself at the simple trade of lens-polishing. He died on Febrary 21, 1677.

Below are excerpts from one of his famous treatises entitled, *The Ethics*.

CONCERNING GOD

Definitions

I. By that which is self-caused, I mean that of which the essence involves existence, or that of which the nature is only conceivable as existent.

II. A thing is called finite after its kind, when it can be limited by another thing of the same nature; for instance, a body is called finite because we always conceive another greater body. So, also, a thought is limited by another thought, but a body is not limited by thought, nor a thought by body.

III. By substance, I mean that which is in itself, and is conceived through itself; in other words, that of which a conception can be formed independently of any other conception.

IV. By attribute, I mean that which the intellect perceives as constituting the essence of substance.

V. By mode, I mean the modifications of substance, or that which exists in, and is perceived through, something other than itself.

VI. By God, I mean a being absolutely infinite—that is, a substance consisting in infinite attributes, of which each expresses eternal and infinite essentiality.

Explanation. I say absolutely infinite, not infinite after its kind; for, of a thing infinite only after its kind, infinite attributes may be denied; but that which is absolutely infinite, contains in its essence whatever expresses reality, and involves no negation.

VII. That thing is called free, which exists solely by the necessity of its own nature, and of which the action is determined by itself alone. On the other hand, that thing is necessary, or rather constrained, which is determined by something external to itself to a fixed and definite method of existence or action.

VIII. By eternity, I mean existence itself, insofar as it is conceived necessarily to follow solely from the definition of that which is external.

Explanation. Existence of this kind is conceived as an eternal truth, like the essence of a thing, and, therefore, cannot be explained by means of continuance or time, though continuance may be conceived without a beginning or end.

Axioms

I. Everything which exists, exists either in itself or in something else.

II. That which cannot be conceived through anything else must be conceived through itself.

III. From a given definite cause, an effect necessarily follows; and, on the other hand, if no definite cause be granted, it is impossible that an effect can follow.

IV. The knowledge of an effect depends on and involves the knowledge of a cause.

V. Things which have nothing in common cannot be understood, the one by means of the other; the conception of one does not involve the conception of the other.

VI. A true idea must correspond with its ideate or object.

VII. If a thing can be conceived as non-existing, its essence does not involve existence.

Propositions

Prop. I. Substance is by nature prior to its modifications.

Proof. This is clear from Def. iii, and v.

Prop. II. Two substances, whose attributes are different, have nothing in common.

Proof. Also evident from Def. iii. For each must exist in itself, and be conceived through itself; in other words, the conception of one does not imply the conception of the other.

Prop. III. Things which have nothing in common cannot be one the cause of the other.

Proof. If they have nothing in common, it follows that one cannot be apprehended by means of the other (Ax. v.), and, therefore, one cannot be the cause of the other (Ax. iv.) Q.E.D.

Prop. IV. Two or more distinct things are distinguished one from the other, either by the difference of the attributes of the substances, or by the difference of their modifications.

Proof. Everything which exists, exists either in itself or in something else (Ax. i.); that is (by Def. iii and v.), nothing is granted in addition to the understanding, except substance and its modifications. Nothing is, therefore, given besides the understanding, by which several things may be distinguished one from the other, except the substances, or, in other words (see Ax. iv.), their attributes and modifications. Q.E.D.

Prop. V. There cannot exist in the universe two or more substances having the same nature or attribute.

Proof. If several distinct substances be granted, they must be distinguished one from the other, either by the difference of their attributes, or by the difference of their modifications (Prop. iv.). If only by the difference of their attributes it will be granted that there cannot be more than one, with an identical attribute. If by the difference of their modifications—as substance is naturally prior to its modifications (Prop. i.)—it follows that setting the modifications aside, and considering substance in itself, that is, truly (Def. iii. and vi.), there cannot be conceived one substance different from another—that is (by Prop. iv.)—there cannot be granted several substances, but one substance only. Q.E.D.

Prop. VI. One substance cannot be produced by another substance.

Proof. It is impossible that there should be in the universe two substances with an identical attribute, i. e., which have anything common to them both (Prop. ii.), and, therefore (Prop. iii.), one cannot be the cause of another, neither can one be produced by the other. Q.E.D.

Corollary. Hence it follows that a substance cannot be produced by anything external to itself. For in the universe nothing is granted, save substances and their modifications (as appears from Ax. i. and Def. iii. and v.). Now (by the last Prop.) substance cannot be produced by anything external to itself. Q.E.D. This is shown still more readily by the absurdity of the contradictory. For, if substance be produced by an external cause, the knowledge of it would depend on the knowledge of its cause (Ax. iv.), and (by Def. iii.) it would itself not be substance.

Prop. VII. Existence belongs to the nature of substance.

Proof. Substance cannot be produced by anything external (Corollary, Prop. vi.), it must, therefore, be its own cause, that is, its essence necessarily involves existence, or existence belongs to its nature.

CHAPTER 35

JOHN LOCKE

JOHN LOCKE WAS BORN in Wrington, Somersetshire, England, in 1632. He obtained his early education at Westminster School in London, and attended Oxford from 1651 to 1664. He became a member of the household of Lord Ashley, who later became first earl of Shaftesbury, and through those connections became greatly interested in politics. He became a leader in political thought. The Civil Government, which he advocated, was entirely too liberal to please King James, and Locke was compelled to retreat to the continent where be lived for five years. In 1690 he published his *Essay Concerning Human Understanding*. The book really begins descriptive psychology. It is mostly confined to an analysis of the nature of knowledge—what knowledge consists of; in other words, his book treated with the subject of epistemology. According to Locke, reality was entirely different from our conception of it. Our idea of reality, or our knowledge of reality, depended upon sensations received and upon our reflection upon these sensations and from our combining of them arose all our knowledge of the exterior world in which we live, and also accounts for our process of reasoning.

He also beautifully and forcefully sets forth this profound subject under the title of *Of Ideas* and it is our pleasure to give you excerpts from his treatise below. It is advisable that you carefully and studiously read these excerpts. You will find them very, very beneficial, and furthermore, you will find them so fundamental that modern-day psychology and modern philosophy deviate very slightly from this earlier conception of what constitutes human knowledge of understanding.

Locke died October 28, 1704.

OF IDEAS IN GENERAL AND THEIR ORIGINAL

Idea is the object of thinking:—Every man being conscious to himself that he thinks; and that which his mind is applied about whilst thinking being the *ideas* that are there, it is past doubt that men have in their minds several ideas, such as are those expressed by the words *whiteness, hardness, sweetness, thinking, motion, man, elephant, army, drunkenness,* and others: it is in the first place then to be inquired, *How he comes by them?* I know it is a received doctrine that men have native ideas and original characters stamped upon their minds in their very first being. This opinion I have at large examined already; and, I suppose what I have said in the foregoing book will be much more easily admitted, when I have shown whence the understanding may get all the ideas it has; and by what ways and degrees they may come into the mind; for which I shall appeal to every one's own observation and experience.

All ideas come from sensation or reflection.—Let us then suppose the mind to be, as we say, white paper, void of all characters, without any ideas: How comes it to be furnished? Whence comes it by that vast store which the busy and boundless fancy of man has painted on it with an almost endless variety? Whence has it all the *materials* of reason and knowledge? To this I answer, in one word, from *Experience.* In that all our knowledge is founded; and from that it ultimately derives itself. Our observation employed either, about external sensible objects, or about the internal operations of our minds, perceived and reflected on by ourselves is that which supplies our understandings with all the *materials* of thinking. These two are the fountains of knowledge, from whence all the ideas we have, or can naturally have, do spring.

The objects of sensation one source of ideas.—First, our Senses, conversant about particular sensible objects, do convey into the mind several distinct perceptions of things, according to those various ways wherein those objects do affect them. And thus we come by those *ideas* we have of *yellow, white, heat, cold, soft, hard, bitter, sweet,* and all those which we call sensible qualities; which when I say the senses convey into the mind, I mean, they from external objects convey into the mind what produces there those perceptions. This great source of most of the ideas we have, depending wholly upon our senses and derived by them to the understanding, I call SENSATION.

The operations of our minds, the other source of them.—Secondly, the other fountain, from which experience furnisheth the understanding with ideas is,—the perception of the operations of our own mind within us, as it is employed about the ideas it has got;—which operations, when the soul comes to reflect on and consider, do furnish the understanding with another set of ideas, which could not be had from things without. And such are *perception, thinking, doubting, believing, reasoning, knowing, willing,* and all the different actings of our own minds;—which we being conscious of, and observing in ourselves, do from these receive into our understandings as distinct ideas as we do from bodies affecting our senses. This source of ideas every man has wholly in himself; and though it be not sense, as having nothing to do with external objects, yet it is very like it, and might properly enough be called *internal sense.* But as I call the other Sensation, so I call this *Reflection*, the ideas it affords being such only as the mind gets by reflecting on its own operations within itself. By reflection then, in the following part of this discourse, I would be understood to mean that notice which the mind takes of its own operations, and the manner of them, by reason whereof there come to be ideas of these operations in the understanding. These two, I say, viz. external material things, as the objects of *Sensation*, and the operations of our own minds within, as the objects of *Reflection*, are to me the only originals from whence all our ideas take their beginnings. The term *operations* here I use in a large sense, as comprehending not barely the actions of the mind about its ideas, but some sort of passions arising sometimes from them, such as is the satisfaction or uneasiness arising from any thought.

All our ideas are of the one or the other of these.—The understanding seems to me not to have the least glimmering of any ideas which it doth not receive from one of these two. *External objects* furnish the mind with the ideas of sensible qualities, which are all those different perceptions they produce in us; and *the mind* furnishes the understanding with ideas of its own operations.

These, when we have taken a full survey of them, and their several modes, combinations, and relations, we shall find to contain all our whole stock of ideas; and that we have nothing in our minds which did not come in one of these two ways. Let any one examine his own thoughts, and thoroughly search into his understanding; and then let him tell me, whether all the original ideas he has there, are any other than of the objects of his senses, or of the operations of his mind, considered as objects of his reflection. And how great a mass of knowledge soever he imagines to be lodged there, he will, upon taking a strict view, see that he has not any idea in his mind but what one of these two have imprinted;—though perhaps, with infinite variety compounded and enlarged by the understanding, as we shall see hereafter.

Observable in children.—He that attentively considers the state of a child, at his first coming into the world, will have little reason to think him stored with plenty of ideas, that are to be the matter of his future knowledge. It is *by degrees* he comes to be furnished with them. And though the ideas of obvious and familiar qualities imprint themselves before the memory begins to keep a register of time or order, yet it is often so late before some unusual qualities come in the way, that there are few men that cannot recollect the beginning of their acquaintance with them. And if it were worth while, no doubt a child might be so ordered as to have but a very few, even of the ordinary ideas, till he were grown up to a man. But all that are born into the world, being surrounded with bodies that perpetually and diversely affect them, variety of ideas, whether care be taken of it or not, are imprinted on the minds of children. Light and colours are busy at hand everywhere, when the eye is but open; sounds and some tangible qualities fail not to solicit their proper senses, and force an entrance to the mind; but yet, I think, it will be granted easily, that if a child were kept in a place where he never saw any other but black and white till he were a man, he would have no more ideas of scarlet or green, than he that from his childhood never tasted an oyster or a pineapple, has of those particular relishes.

Men are differently furnished with these, according to the different objects they converse with.—Men then come to be furnished with fewer or more simple ideas from without, according as the objects they converse with afford greater or less variety; and from the operations of their minds within, according as they more or less reflect on them. For though he that contemplate the operations of his mind, cannot but have plain and clear ideas of them; yet, unless he turns his thoughts that way, and considers them *attentively*, he will no more have clear and distinct ideas of all the operations of his mind, and all that may be observed therein, than he will have all the particular ideas of any landscape, or of the parts and motions of a clock, who will not turn his eyes to it, and with attention heed all the parts of it. The picture or clock may be so placed, that they may come in his way every day; but yet he will have but a confused idea of all the parts they are made up of, till he applies himself with attention to consider them each in particular.

CHAPTER 36

NICOLAS MALEBRANCHE

THE EARLY LIFE OF this mystic and philosopher is an example of how we may be pursuing a wrong course in life and yet, by a fortunate combination of circumstances, arouse a dormant desire which eventually leads us to success and attainment. It is indeed regrettable when such circumstances do not occur in the life of one who is following a channel foreign to his interests.

Nicolas Malebranche, a French philosopher, was born in Paris, August 6, 1638, and he died October 13, 1715. He came from a prominent family and was the youngest child of Nicolas Malebranche, Secretary to Louis XIII. He had the advantage of an excellent education and began his studies at the Collège of La March, later studied theology at the Sorbonne. It was his early intention to enter the church, and he was ordained, but his love of retirement led him to decline a Canonicate in Notre Dame.

When still a young man, in fact, 22 years of age, he entered the Congregation of the Oratory and devoted himself to the study of ecclesiastical history. He found it extremely difficult, however, to harmonize the various incidents, and was losing interest in his studies, when he came across Descartes' *Traité de l'Homme*, which aroused a dormant enthusiasm for philosophy. He heartily agreed with Descartes' distinction between mind and matter, and considered the only true qualities of matter, extension, and motion. Malebranche is particularly renowned for his work entitled *Recherche de la Verité.*

We bring to you, below, excerpts from this writing which he entitled *What Is Meant By Ideas.* It is greatly condensed, yet sufficient to show you the depth of his thoughts and from a philosophical point of view is today equally as instructive and interesting as when written.

WHAT IS MEANT BY IDEAS

I suppose that everyone will grant that we perceive not the objects that are without us immediately and of themselves. We see the sun, the stars, and infinite other objects without us; and it is not probable that the soul goes out of the body, and fetches a walk, as I may say, about the heavens, to contemplate all the objects therein.

It sees them not therefore by themselves, and the immediate object of the mind, when it beholds the sun, for example, is not the sun, but something intimately united to the soul; and that same thing is what I call our "idea." So that by the term idea I mean nothing but that object which is immediate, or next, to the soul in its perception of anything.

It ought to be well observed that in order to the mind's perceiving any object it is absolutely necessary the idea of that object be actually present to it: which is so certain as not possibly to be doubted of. But it is not necessary there should be anything without like to that idea; for it often happens that we perceive things which do not exist, and which never were in nature. And so a man has frequently in his mind real ideas of things that never were. When a man, for instance, imagines a golden mountain, it is indispensably necessary that the idea of that mountain should be really present in his mind. When a frantic, or a man in a fever or sleep, sees some terrible animal before his eyes, it is certain that the idea of that animal really exists. And yet that mountain of gold and this animal never were in being.

Notwithstanding, men being, as it were, naturally inclined to believe that corporeal objects exist, judge of the reality and existence of things quite otherwise than they ought. For when they perceive an object by way of sense, they will have it most infallibly to exist, though it often happens that there is nothing of it without; they will have, moreover, this object to be just the same as they perceive it; which yet never happens. But as for the idea which necessarily exists, and cannot be otherwise than we see it, they commonly judge, without reflection, that it is nothing at all: as if ideas had not a vast number of properties (as that the idea of a square, for instance, were not very different from that of any number), and did not represent quite different things! Which is not consistent with nothing, since nothing has no property. It is therefore undoubtedly certain that ideas have a most real existence. But let us inquire into their nature and their essence, and see what there is in our soul capable of making to her the representations of all things.

Whatever things the soul perceives are only of two sorts, and are either within or without the soul. Those that are within the soul are its own proper thoughts; that is, all its different modifications. For by the words "thought," "manner of thinking," or "modifications of the soul," I mean all those things in general which cannot be in the soul without her perceiving them; such are her own sensations, her imaginations, her pure intellections, or simply her conceptions, as also her passions and natural inclinations. Now our soul has no need of ideas to perceive all these things, because they are within the soul, or, rather, because they are the very soul itself, in such or such a manner: just as the real rotundity of any body and its motion are nothing but the body figured and translated, after such or such a sort.

But as to the things without the soul, we can have no perception of them but by the means of ideas, upon supposition that these things cannot be intimately united to it; and they are of two sorts, *Spiritual* and *Material*; as to the Spiritual, there is some probability they may be discovered to the soul without ideas, immediately by themselves. For though experience certifies us that we cannot by an immediate communication, declare our thoughts to one another, but only by words and other sensible signs whereunto we have annexed our ideas; yet we may say that God has ordained this kind of economy only for the time of this life, to prevent the disorders that might at present happen if men should understand one another as they pleased. But when justice and order shall reign, and we shall be delivered from the captivity of our body, we shall possibly communicate our thoughts by the intimate union of ourselves, as it is probable the angels may do in heaven. So that there seems to be no absolute necessity of admitting ideas for the representing things of a spiritual nature, since it is possible for them to be seen by themselves, though in a very dark and imperfect manner.

CHAPTER 37

SIR ISAAC NEWTON

IT IS INDEED UNFORTUNATE that fantastic tales, undoubtedly in many instances unfounded in fact, are attributed to eminent characters of history, and which are said to account for their great discoveries or contributions to the world. Such myths really do them an injustice inasmuch as a true investigation of their lives usually reveals their discoveries were born of great mental labor and much profound thought.

Sir Isaac Newton, born on December 25, 1642, at Woolsthorpe, Lincolnshire, England, is one such eminent contributor to the world's knowledge whose great work has the setting of a tale of a falling apple upon his head as the cause of his resulting scientific ideas. Popularly he is known as the discoverer of the Law of Gravity and believed to have won eminence by that particular contribution alone. If he had never made that discovery, his other contributions would have been sufficient to have made him an immortal. His beginnings were quite humble. He attended the Free Grammar School in Grantham, but left early. His ability to make mechanical toys was the cause of his parents' returning him to school and later sending him to Cambridge.

He acquired a degree in 1665, and in 1667 was made a fellow of the university. Two years later he held a chair in mathematics. He made startling discoveries in mathematics, which in and by themselves were sufficient to make him renowned. In 1672 he was admitted to membership in the Royal Society. A little later he began his famous experiments with light, proving that white light is a compound of various color rays. He advanced the knowledge of the laws of optics. He finally conceived the idea of a universal gravity. He calculated the pull the Earth would exert on the Moon in accordance with its supposed mass. However, his hypothesis did not agree with the speed of revolutions of the Moon, and he temporarily laid it aside.

In 1682 a new measurement of the meridian was brought to his attention and revived his interest in the hypothesis. To his great pleasure his new calculations worked out satisfactorily, and he expounded a universal law for attraction between cosmic bodies.

Later in life he received political distinction by being appointed warden of the mint between 1695 and 1699 at an excellent annual salary, which made it possible to equip his home laboratory for further research.

It may interest the reader to know that we have in the archives of the Rosicrucian Order the reproduction of a flyleaf from a book by Sir Isaac Newton upon which is written in his own handwriting, signed and dated by himself, comments on the activities and history of the Rosicrucian Order, identifying his interest with the Rosicrucian activities of his period.

THE DIFFUSION OF LIGHT

A letter of Mr. Isaac Newton, Professor of Mathematics in the University of Cambridge; containing his new theory of Light and Colours; sent by the Author to the Editor from Cambridge, Feb. 6, 1671-3; to be communicated to the Royal Society. No. 80, p. 3,075.

Sir—to perform my late promise to you, I shall without further ceremony acquaint you that in the beginning of the year 1666 (at which time I applied myself to the grinding of optic glasses of other figures than spherical,) I procured a triangular glass prism, to try therewith the celebrated phenomena of colours. And for that purpose, having darkened my chamber, and made a small hole in my window shuts, to let in a convenient quantity of the sun's light, I placed my prism at this entrance, that it might be thereby refracted to the opposite wall. It was at first a very pleasing diversion to view the vivid and intense colours produced thereby; but after a while applying myself more circumspectly, I was surprised to see them in an oblong form; which according to the received laws of refraction, I expected would have been circular. They were terminated at the sides with straight lines, but at the ends the decay of light was so gradual, that it was difficult to determine justly what was their figure; yet they seemed semicircular.

Comparing the length of this coloured spectrum with its breadth, I found it about five times greater; a disproportion so extravagant, that it excited me to a more than ordinary curiosity of examining from whence it might proceed. I could scarce think that the various thickness of the glass, or the termination with shadow or darkness, could have any influence on light to produce such an effect; yet I thought it not amiss, first to examine those circumstances, and so tried what would happen by transmitting light through parts of the glass of divers thicknesses, or through holes in the window of divers sizes, or by setting the prism without, so that the light might pass through it, and be refracted before it was terminated by the hole; but I found none of these circumstances material. The fashion of the colours was in all these cases the same.

Then I suspected, whether by any unevenness in the glass, or other contingent irregularity, these colours might be thus dilated. And to try this, I took another prism like the former, and so placed it, that the light passing through them both, might be refracted contrary ways, and so be the latter returned into the course from which the former had diverted it. For, by this means, I thought the regular effects of the first prism would be destroyed by the second, but the irregular ones more augmented, by the multiplicity of refractions. The event was, that the light, which by the first prism was diffused into an oblong form, was by the second reduced into an orbicular one, with as much regularity as when it did not at all pass through them. So that, whatever was the cause of that length, it was not any contingent irregularity.

I then proceeded to examine more critically, what might be effected by the difference of the incidence of rays coming from divers parts of the sun; and to that end measured the several lines and angles, belonging to the image. Its distance from the hole or prism was 22 feet; its utmost length 13¼ inches; its breadth 2⅝; the diameter of the hole ¼ of an inch; the angle, which the rays, tending towards the middle of the image, made with those lines in which they would have proceeded without refraction, was 44° 56′. And the vertical angle of the prism, 63° 12′. Also the refraction on both sides the prism, that is of the incident and emergent rays, was as near as I could make them equal, and consequently about 54° 4′. And the rays fell perpendicularly upon the wall. Now subducting the diameter of the hole from the length and breadth of the image, there remains 13 inches the length, and 2⅜ the breadth, comprehended, by those rays, which passed through the center of the said hole, and consequently the angle of the hole, which that breadth subtended, was about 31′, answerable to the sun's diameter; but the angle which its length subtended, was more than five such diameters, namely 2° 49′.

Having made these observations, I first computed from them the refractive power of that glass, and found it measured by the ratio of the sizes, 20 to 31. And then, by the ratio, I computed the refraction of two rays flowing from opposite parts of the sun's discus, so as to differ 31′ in their obliquity of incidence, and found that the emergent rays should have comprehended an angle of about 31′, as they did, before they were incident. But because this computation was founded on the hypothesis of the proportionality of the sines of incidence and refraction, which though, by own experience, I could not imagine to be so erroneous as to make that angle but 31′, which in reality was 2° 49′; yet my curiosity caused me again to take my prism. And having placed it at my window, as before, I observed, that by turning it a little about its axis to and fro, so as to vary its obliquity to the light, more than an angle of 4 or 5 degrees, the colours were not thereby sensibly translated from their place on the wall, and consequently by that variation of incidence, the quantity of refraction was not sensibly varied. By this experiment, therefore, as well as by former computation, it was evident, that the difference of the incidence of rays, flowing from divers parts of the sun, could not make them after a decussion, diverge at a sensibly greater angle, than that at which they before converged, which begin at most but about 21 or 32 minutes, there still remained some other cause to be found out, from whence it could be 2° 49′.

Then I began to suspect whether the rays, after their trajection through the prism, did not move in curve lines, and according to their more or less curvity tend to divers parts of the wall. And it increased my suspicion, when I remembered that I had often seen a tennis ball, struck with an oblique racket, describe such a curved line. For, a circular as well as a progressive motion being communicated to it by that stroke, its parts on that side, where the motions conspire, must press and beat the contiguous air more violently than on the other, and there excite a reluctancy and reaction of the air proportionately greater. And for the same reason, if the rays of light should possibly be globular bodies, and by their oblique passage out of one medium into another acquire a circulating motion, they ought to feel the greater resistance from the ambient aether, on that side where the motions conspire, and thence be continually bowed to the other. But notwithstanding this plausible ground of suspicion, when I came to examine it, I could observe no such curvity in them. And besides (which was enough for my purpose) I observed, that the difference between the length of the image and the diameter of the hole, through which the light was transmitted, was proportionable to their distance.

The gradual removal of these suspicions at length led me to the experimentum crucis, which was this; I took two boards, and placed one of them close behind the prism at the window, so that the light might pass through a small hole, made in it for that purpose, and fall on the other board, which I placed at about 12 feet distance, having first made a small hole in it also, for some of that incident light to pass through. Then I placed another prism behind this second board, so that the light trajected through both of the boards, might pass through that also, and be again refracted before it arrived at the wall. This done, I took the first prism in my hand, and turned it to and fro slowly about its axis, so much as to make the several parts of the image, cast on the second board, successively pass through the hole in it, that I might observe to what places on the wall the second prism would refract them. And I saw, by the variation of those places, that the light tending to that end of the image, towards which the refraction of the first prism was made, did in the second prism suffer a contraction considerably greater than the light tending to the other end. And so the true cause of the length of that image was detected to be no other, than that light consists of rays differently refrangible, which, without any respect to a difference in their incidence, were according to their degrees of refrangibility, transmitted towards divers parts of the wall. . . .

Hence it therefore comes to pass, that whiteness is the usual colour of light: for, light is a confused aggregate of rays imbued with all sorts of colours, as they are promiscuously darted from the various parts of luminous bodies. And of such a confused aggregate, as I said, is generated whiteness, if there be a due proportion of the ingredients, but if any one predominate, the light must incline to that colour; as it happens in the blue flame of brimstone; the yellow flame of a candle; and the various colours of the fixed stars.

These things considered, the manner how colours are produced by the prism is evident. For, of the rays constituting the incident light since those which differ in colour, proportionally differ in refrangibility, they by their unequal refractions must be severed and dispersed into an oblong form in an orderly succession, from the least refracted scarlet, to the most refracted violet. And for the same reason it is that objects, when looked upon through a prism, appear coloured. For the difform rays, by their unequal refractions, are made to diverge towards several parts of the retina, and there express the images of things coloured, as in the former case they did the sun's image upon the wall. And by this inequality of refractions they became not only coloured, but also very confused and indistinct.

Why the colours of the rainbow appear in falling drops of rain, is also from hence evident. For, those drops which refract the rays disposed to appear purple, in greatest quantity to the spectator's eye, refract those of other sorts so much more, as to make them pass beside it; and such are the drops on the exterior part of the primary, and interior part of the secondary bow.

The old phenomena of an infusion of lignum nephriticum, leaf gold, fragments of coloured glass, and some other transparently coloured bodies, appearing in one position of one colour, and of another in another, are on these grounds no longer riddles. For, those are substances apt to reflect one sort of light, and transmit another; as may be seen in a dark room, by illuminating them with similar or uncompounded light. For, then they appear that colour only, with which they are illuminated, but yet in one position more vivid and luminous than in another, accordingly as they are disposed more or less to reflect or transmit the incident colour.

From hence also is manifest the reason of an unexpected experiment, which Mr. Hook, somewhere in his micography, relates to have made with two wedge-like transparent vessels, filled the one with red, the other with a blue liquor: namely, that though they were severally transparent enough, yet both together became opaque; for, if one transmitted only red, and the other only blue, no rays could pass through them both.

I might add more instances of this nature; but I shall conclude with this general one, that the colours of all natural bodies have no other origin than this, that they are variously qualified to reflect one sort of light in greater plenty than another. And this I have experimented in a dark room, by illuminating those bodies with uncompounded light of divers colours. For, by that means, any body may be made to appear of any colour. They have then no appropriate colour, but ever appear of the colour of the light cast upon them, but yet with this difference, that they are most brisk and vivid in the light of their own daylight colour. Minium appears there of any colour indifferently, with which it is illustrated, but yet most luminous is red; and so bise appears indifferently of any colour with which it is illustrated, but yet most luminous in blue. And therefore minium reflects rays of any colour, but most copiously those indued with red: and consequently when illustrated with daylight, that is, with all sorts of rays promiscuously blended, those qualified with red shall abound most in the reflected light, and by their prevalence cause it to appear of that colour. And for the same reason bise, reflecting blue most copiously, shall appear blue by the excess of those rays in its reflected light; and the like of other bodies. And that this is the entire and adequate cause of their colours, is manifest, because they have no power to change or alter the colours of any sort of rays, incident apart, but put on all colours indifferently, with which they are enlightened.

CHAPTER 38

WILLIAM PENN

WE WILL DEVOTE OUR attention next to William Penn, who was born in London, October 14, 1644. He had received most of his schooling at Christ Church, Oxford, until his ardent enthusiasm for the Quaker belief and his evident dislike for the Church of England prompted extreme actions on his part and caused him to be expelled. His father hoped that travel on the Continent would help to change his viewpoints, but he clung tenaciously to his faith. He studied law and later became a minister, preaching and writing treatises which often caused his imprisonment. His most famous treatise was written while in prison in vindication of his beliefs and was entitled *No Cross, No Crown.*

Circumstances turned his attention to the New World, however, and in 1681 Penn obtained from the crown a grant of territory which is now the state of Pennsylvania, and in doing this realized his great ambition to establish in America a home for his coreligionists where they might preach and practice their convictions unmolested. In 1682 he founded the capital which he called Philadelphia, a name beloved to all Rosicrucians. It was the location of the first Rosicrucian Headquarters in America in 1694. These pioneer Rosicrucians were sponsored by the Philadelphians in England. Not only Quakers but persecuted members of other religions sought refuge in his colony where, from the first, the principles of toleration were established by law. Penn made himself famous by his honest negotiations with the Indians, which resulted in a lasting treaty of peace. His colony thriving successfully, he then returned to England and through his influence obtained the release of hundreds of persons who were imprisoned on account of their religious beliefs. The latter part of his life was involved in periodic charges of treason and conspiracy against the crown, acquittals, and finally his financial ruin because of the treachery of his steward. Later he was stricken with paralysis and passed through transition May 30, 1718.

REFLECTIONS AND MAXIMS

It is admirable to consider how many millions of people come into and go out of the world, ignorant of themselves, and of the world they have lived in.

If one went to see Windsor-castle, or Hampton-court, it would be strange not to observe and remember the situation, the building, the gardens, fountains, etc., that make up the beauty and pleasure of such a seat. And yet few people know themselves; no, not their own bodies, the houses of their minds, the most curious structure in the world; a living, walking tabernacle; nor the world of which it was made, and out of which it is fed; which would be so much our benefit, as well as our pleasure, to know. We cannot doubt of this when we are told that the "invisible things of God are brought to light by the things that are seen;" and consequently we read our duty in them, as often as we look upon them, to him that is the great and wise author of them, if we look as we should do.

The world is certainly a great and stately volume of natural things, and may be not improperly stiled the hieroglyphics of a better; but, alas, how very few leaves of it do we seriously turn over! This ought to be the subject of the education of our youth; who, at twenty, when they should be fit for business, know little or nothing of it.

Inconsideration

The want of due consideration is the cause of all the unhappiness man brings upon himself. For his second thoughts rarely agree with the first; which pass not without considerable retrenchment or correction. And yet that sensible warning is, too frequently, not precaution enough for his future conduct.

Well may we say, "Our infelicity is of ourselves;" since there is nothing we do that we should not do, but we know it, and yet do it.

Disappointment and Resignation

For disappointments, that come not by our own folly, they are the trials or corrections of heaven; and it is our own fault, if they prove not our advantage.

To repine at them does not mend the matter; it is only to grumble at our Creator. But to see the hand of God in them, with an humble submission to his will, is the way to turn our water into wine, and engage the greatest love and mercy on our side.

We needs must disorder ourselves, if we only look at our losses. But if we consider how little we deserve what is left, our passion will cool, and our murmurs will turn into thankfulness.

If our hairs fall not to the ground, less do we, or our substance, without God's providence.

Nor can we fall below the arms of God, how low soever it be we fall.

For though our Saviour's passion is over, his compassion is not. That never fails his humble, sincere disciples. In him they find more than all that they lose in the world.

Discipline

If thou wouldst be happy and easy in thy family, above all things observe discipline.

Every one in it should know their duty; and there should be a time and place for everything; and, whatever else is done or omitted, be sure to begin and end with God.

Industry

Love labour: for if thou does not want it for food, thou mayst for physic. It is wholesome for thy body, and good for thy mind. It prevents the fruits of idleness, which many times come of nothing to do, and lead too many to do what is worse than nothing.

A garden, a laboratory, a workhouse, improvements, and breeding, are pleasant and profitable diversions to the idle and ingenious; for here they miss ill company, and converse with nature and art; whose varieties are equally grateful and instructing, and preserve a good constitution of body and mind.

Temperance

To this a spare diet contributes much. Eat therefore to live, and do not live to eat. That is like a man, but this below a beast.

Have wholesome, but not costly food; and be rather cleanly than dainty in ordering it.

The receipts of cookery are swelled to a volume, but a good stomach excels them all; to which nothing contributes more than industry and temperance.

It is a cruel folly to offer up to ostentation so many lives of creatures, as to make up the state of our treats; as it is a prodigal one to spend more in sauce than in meat.

The proverb says, "That enough is as good as a feast;" but it is certainly better, if superfluity be a fault, which never fails to be at festivals.

If thou rise with an appetite, thou are sure never to sit down without one.

Rarely drink but when thou art dry; nor then, between meals, if it can be avoided.

The smaller the drink, the clearer the head, and the cooler the blood, which are great benefits in temper and business.

Strong liquors are good at some times, and in small proportions; being better for physic than food; for cordials, than common use.

The most common things are the most useful; which shews both the wisdom and goodness of the great Lord of the family of the world.

What, therefore, he has made rare, do not thou use too commonly; lest thou shouldst invert the use and order of things, become wanton and voluptuous, and thy blessings prove a curse.

"Let nothing be lost," said our Saviour; but that is lost that is misused.

Neither urge another to that thou wouldst be unwilling to do thyself; nor do thyself what looks to thee unseemly and intemperate in another.

All excess is ill; but drunkenness is of the worst sort. It spoils health, dismounts the mind, and unmans men. It reveals secrets, is quarrelsome, lascivious, impudent, dangerous, and mad. In fine, he that is drunk is not man; because he is so long void of reason, that distinguishes a man from a beast.

Friendship

Friendship is the next pleasure we may hope for; and where we find it not at home, or have no home to find it in, we may seek it abroad. It is an union of spirits, a marriage of hearts, and the bond thereof virtue.

There can be no friendship where there is no freedom. Friendship loves a free air, and will not be penned up in straight and narrow inclosures. It will speak freely, and act so too; and take nothing ill, where no ill is meant; nay, where it is, it will easily forgive, and forget, upon small acknowledgments.

Friends are true twins in soul; they sympathize in every thing, and have the same love and aversion.

One is not happy without the other; nor can either of them be miserable alone. As if they could change bodies, they take their turns in pain as well as in pleasure; relieving one another in their most adverse conditions.

What one enjoys, the other cannot want. Like the primitive Christians, they have all things in common, and no property, but in one another.

Rules of Conversation

Avoid company, where it is not profitable or necessary; and on those occasions speak little, and last.

Silence is wisdom where speaking is folly, and always safe.

Some are so foolish, as to interrupt and anticipate, those that speak, instead of hearing and thinking before they answer; which is uncivil, as well as silly.

If thou thinkest twice before thou speakest once, thou wilt speak twice the better for it.

Better say nothing, than not to the purpose. And to speak pertinently, consider both what is fit, and when it is fit, to speak.

In all debates, let truth be thy aim; not victory, or an unjust interest; and endeavour to gain, rather than to expose, thy antagonist.

Give no advantage in argument, nor lose any that is offered. This is a benefit which arises from temper.

Do not use thyself to dispute against thine own judgment, to shew wit; lest it prepare thee to be too indifferent about what is right; nor against another man, to vex him, or for mere trial of skill; since to inform, or to be informed, ought to be the end of all conferences.

Temper

Nothing does reason more right than the coolness of those that offer it; for truth often suffers more by the heat of its defenders, than from the arguments of its opposers.

Zeal ever follows an appearance of truth, and the assured are too apt to be warm; but it is their weak side in argument; zeal being better shewn against sin, than persons, or their mistakes.

Truth

Where thou art obliged to speak, be sure to speak the truth; for equivocation is half-way to lying, as lying the whole way to hell.

Knowledge

Knowledge is the treasure, but judgment the treasurer, of a wise man.

He that has more knowledge than judgment, is made for another man's use more than his own.

It cannot be a good constitution, where the appetite is great, and the digestion weak.

There are some men, like dictionaries, to be looked into upon occasions; but have no connection, and are little entertaining.

Less knowledge than judgment, will always have the advantage upon the injudicious knowing man.

A wise man makes what he learns his own; the other shews he is but a copy, or a collection at most.

Industry

Industry is certainly very commendable, and supplies the want of parts.

Patience and diligence, like faith, remove mountains.

Never give out while there is hope; but hope not beyond reason; for that shews more desire than judgment.

Balance

We must not be concerned above the value of the thing that engages us; nor raised above reason, in maintaining what we think reasonable.

It is too common an error, to invert the order of things, by making an end of that which is a means, and a means of that which is an end.

Religion and government escape not this mischief; the first is too often made a means, instead of an end; the other an end, instead of a means.

Thus men seek wealth, rather than subsistence; and the end of cloaths is the least reason of their use. Nor is the satisfying of our appetite our end in eating, so much as the pleasing of our palate. The like may also be said of building, furniture, etc., where the man rules not the beast, and appetite submits not to reason.

It is great wisdom to proportion our esteem to the nature of the thing; for as that way things will not be undervalued, so neither will they engage us above their intrinsic worth.

If we suffer little things to have great hold upon us, we shall be as much transported for them, as if they deserved it.

It is an old proverb, *"Maxima bella ex levissimis causis,"* The greatest feuds have had the smallest beginnings.

No matter what the subject of the dispute be, but what place we give it in our minds; for that governs our concern and resentment.

It is one of the fatalest errors of our lives, when we spoil a good cause by an ill management; and it is not impossible but we may mean well in an ill business; but that will not defend it.

If we are but sure the end is right, we are too apt to gallop over all bounds to compass it; not considering, that lawful ends may be very unlawfully attained.

Let us be careful to take just ways to compass just things; that they may last in their benefits to us.

There is a troublesome humour some men have, that if they may not lead, they will not follow; but had rather a thing were never done, than not done their own way, though otherwise very desirable.

This comes of an overfulness of ourselves, and shows we are more conerned for praise, than the success of what we think a good thing.

Patience

Patience is a virtue everywhere; but it shines with greatest lustre in the men of government.

Some are so proud or testy, they will not hear what they should redress.

Others so weak, they sink, or burst, under the weight of their office; though they can lightly run away with the salary of it.

Business can never be well done, that is not well understood; which cannot be without patience.

It is cruelty, indeed, not to give the unhappy an hearing, whom we ought to help; but it is the top of oppression to brow-beat the humble and modest miserable, when they seek relief.

Some, it is true, are unreasonable in their desires and hopes; but then we inform, not rail at and reject them

It is, therefore, as great an instance of wisdom as a man in business can give, to be patient under the impertinencies and contradictions that attend it.

Method goes far to prevent trouble in business; for it makes the task easy, hinders confusion, saves abundance of time, and instructs those that have depending, what to do, and what to hope.

Popularity

Affect not to be seen, and men will less see thy weakness.

They that show more than they are, raise an expectation they cannot answer; and so lose their credit, as soon as they are found out.

Avoid popularity. It has many snares, and no real benefit to thyself; and uncertainty to others.

CHAPTER 39

GOTTFRIED WILHELM LEIBNITZ

LEIBNITZ' PHILOSOPHY IS DEFINITELY related to the period known as the beginning of modern philosophy. He was born in Leipzig, June 21, 1646, and received his general education at the University of Leipzig. In 1667 he was invited by Baron von Boineburg to come to Frankfort as councilor to the Elector of Mainz. His scientific inclinations were aroused by several visits to Paris in 1672 and to London in 1673, where he met many leading scientific men. Publicity was given his philosophical views and his mathematical genius when controversy arose between himself and Newton, because of his system of differential calculus which greatly resembled Newton's method of fluxions.

In 1676 he was appointed librarian to the Duke of Brunswick-Luneburg. He also served as councilor and historian. Although he made his home in Hanover, he visited the courts of many monarchs. Among his chief works in philosophy are the *Monadologia* and, in natural science, his *Protogaea*, a treatise on geology. Our readers, especially Rosicrucian readers, will be particularly interested in his metaphysical views, and below are two excerpts from these metaphysical writings which are worthy of the careful attention of every student.

HOW CONCEPTIONS ARE DERIVED FROM THE SENSES

Aristotle preferred to compare our souls to blank tablets prepared for writing, and he maintained that nothing is in the understanding which does not come through the senses. This position is in accord with the popular conceptions as Aristotle's positions usually are. Plato thinks more profoundly. Such tenets or practicologies are nevertheless allowable in ordinary use somewhat in the same way as those who accept the Copernican theory still continue to speak of the rising and setting of the sun. I find indeed that these usages can be given a real meaning containing no error, quite in the same way as I have already pointed out that we may truly say particular substances act upon one another. In this same sense we may say that knowledge is received from without through the medium of the senses because certain exterior things contain or express more particularly the causes which determine us to certain thoughts. Because in the ordinary uses of life we attribute to the soul only that which belongs to it most manifestly and particularly, and there is no advantage in going further. When, however, we are dealing with the exactness of metaphysical truths, it is important to recognize the powers and independence of the soul which extend infinitely further than is commonly supposed. In order, therefore, to avoid misunderstandings it would be well to choose separate terms for the two. These expressions which are in the soul, whether one is conceiving of them or not, may be called ideas, while those which one conceives of or constructs may be called conceptions, *conceptus*. But whatever terms are used, it is always false to say that all our conceptions come from the so-called external senses, because those conceptions which I have of myself and of my thoughts, and consequently of being, of substance, of action, of identity and of many others come from an inner experience.

Maxims

A body is an aggregation of substances and is not a substance, properly speaking. Consequently, in all bodies must be found indivisible substances which cannot be generated and are not corruptible, having something which corresponds to souls.

All these substances have been always and will always be united to organize bodies diversely transformable.

Each of these substances contains in its nature the law of the continuous progression of its own workings and all that has happened to it and all that will happen to it.

Excepting the dependence upon God, all these activities come from its own nature.

Each substance expresses the whole universe, some substances, however, more distinctly than others, each one especially distinctly with regard to certain things and according to its own point of view.

The union of the soul with the body and even the action of one substance upon another consist only in the perfect mutual accord, expressly established by the ordinance of the first creation, by virtue of which each substance following its own laws falls in with what the others require and thus the activities of the one follow or accompany the activities or changes of the other.

Intellects, or souls which are capable of reflection and of knowing the eternal truths and God have many privileges that exempt them from the transformation of bodies.

In regard to them moral laws must be added to physical laws.

The better things are understood, the more are they found beautiful and comfortable to the desires which a wise man might form.

Those who are not content with the ordering of things cannot boast of loving God properly.

Justice is nothing else than love felt by the wise.

Charity is universal benevolence whose fulfillment the wise carry out comformably to the dictates of reason so as to obtain the greatest good.

Wisdom is the science of happiness or of the means of attaining the lasting contentment which consists in the continual achievement of a greater perfection or at least in variations of the same degree of perfection.

CHAPTER 40

GEORGE BERKELEY

GEORGE BERKELEY WAS BORN in Kilkenny, Ireland, in March of the year 1685. He became a student in Trinity College, Dublin, in 1700. He was connected with that college for 24 years. His interest in philosophy was aroused by the philosophical writings of this period, namely, those of Locke and Descartes. He published his *Essay Towards a New Theory of Vision* in 1709, in which he declared that all that we know of the world of reality is color sensation, and that this color sensation does not impart to us space dimensions, or distance, but that these arise within the consciousness. In other words, he infers, as the Rosicrucians have taught for years, that space and time are matters of consciousness, an interpretation of the sensations of matter. He published his *Principles of Human Knowledge* in 1710. In this he further declared that knowledge had no substance of its own, that the things we declare we know have no existence as we realize them, and that they are merely sensations received through the senses. Therefore, all knowledge is purely mental, and ideas are the cognizance of sensations and reflection upon them.

He spent three years in America where he helped to found one university and donated land and books to another. He died at Oxford, England, in 1753.

PRINCIPLES OF HUMAN KNOWLEDGE

It is evident to any one who takes a survey of the objects of human knowledge that they are either *ideas* actually imprinted on the senses; or else such as are perceived by attending to the passions and operations of the mind; or lastly *ideas* formed by help of memory and imagination—either compounding, dividing, or barely representing those originally perceived in the aforesaid ways.—By sight I have the ideas of light and colours, with their several degrees and variations. By touch I perceive hard and soft, heat and cold, motion and resistance, and of all these more and less either as to quantity or degree. Smelling furnishes me with odours; the palate with tastes; and hearing conveys sounds to the mind in all their variety of tone and composition.—And as several of these are observed to accompany each other, they come to be marked by one name, and so to be reputed as one *thing*. Thus, for example, a certain colour, taste, smell, figure and consistence having been observed to go together, are accounted one distinct thing, signified by the name *apple*; other collections of ideas constitute a stone, a tree, a book, and the like sensible things—which as they are pleasing or disagreeable excite the passions of love, hatred, joy, grief, and so forth.

But, besides all that endless variety of ideas or objects of knowledge, there is likewise something which knows or perceives them; and exercises divers operations, as willing, imagining, remembering, about them. This perceiving, active being is what I call *mind, spirit, soul,* or *myself*. By which I do not denote any one of my ideas, but a thing entirely distinct from them, wherein they exist, or, which is the same thing, whereby they are perceived—for the existence of an idea consists in being perceived.

That neither our thoughts, nor passions, nor ideas formed by the imagination, exist without the mind, is what everybody will allow.—And to me it is no less evident that the various *sensations*, or *ideas imprinted on the sense*, however blended or combined together (that is, whatever *objects* they compose), cannot exist otherwise than in a mind perceiving them.—I think an intuitive knowledge may be obtained of this by any one that shall attend to *what is meant by the term exist when applied to sensible things*. The table I write on I say exists, that is, I see and feel it; and if I were out of my study I should say it existed—meaning thereby that if I was in my study I might perceive it, or that some other spirit actually does perceive it. There was an odour, that is, it was smelt; there was a sound, that is, it was heard; a colour or figure, and it was perceived by sight or touch. This is all that I can understand by these and the like expressions.—For as to what is said of the absolute existence of unthinking things without any relation to their being perceived, that is to me perfectly unintelligible. Their *esse* is *percipi*, nor is it possible they should have any existence out of the minds or thinking things which perceive them.

It is indeed an opinion strangely prevailing amongst men, that houses, mountains, rivers, and in a word all sensible objects, have an existence, natural or real, distinct from their being perceived by the understanding. But, with how great an assurance and acquiescence soever this principle may be entertained in the world, yet whoever shall find it in his heart to call it in question may, if I mistake not, perceive it to involve a manifest contradiction. For, what are the forementioned objects but the things we perceive by sense? and what do we perceive besides our own ideas or sensations? and is it not plainly repugnant that any one of *these*, or any combination of them, should exist unperceived?

If we thoroughly examine this tenet it will, perhaps, be found at bottom to depend on the doctrine of *abstract ideas*. For can there be a nicer strain of abstraction than to distinguish the *existence* of sensible objects from their *being perceived*, so as to conceive them existing unperceived? Light and colours, heat and cold, extension and figures—in a word the things we see and feel—what are they but so many sensations, notions, ideas, or impressions on the sense, and is it possible to separate, even in thought, any of these from perception? For my part, I might as easily divide a thing from itself. I may, indeed, divide in my thoughts, or conceive apart from each other, those things which, perhaps, I never perceived by sense so divided. Thus, I imagine the trunk of a human body without the limbs, or conceive the smell of a rose without thinking of the rose itself. So far, I will not deny, I can abstract—if that may properly be called *abstraction* which extends only to the conceiving separately such objects as it is possible may really exist or be actually perceived asunder. But my conceiving or imagining power does not extend beyond the possibility of real existence or perception. Hence, as it is impossible for me to see or feel anything without an actual sensation of that thing, so it is impossible for me to conceive in my thoughts any sensible thing or object distinct from the sensation or perception of it. (In truth, the object and the sensation are the same thing and cannot, therefore, be abstracted from each other.)

Some truths there are so near and obvious to the mind that a man need only open his eyes to see them. Such I take this important one to be, viz., that all the choir of heaven and furniture of the earth, in a word all those bodies which compose the mighty frame of the world, have not any subsistence without a mind—that their *being* is to *be perceived or known*; that consequently so long as they are not actually perceived by me, or do not exist in my mind or that of any other created spirit, they must either have no existence at all, or else subsist in the mind of some Eternal Spirit—it being perfectly unintelligible, and involving all the absurdity of abstraction, to attribute to any single part of them an existence independent of a spirit. To be convinced of which, the reader need only reflect, and try to separate in his own thoughts the *being* of a sensible thing from its *being perceived*.

From what has been said it is evident there is not any other Substance than *spirit*, or *that which perceives*. But, for the fuller demonstration of this point, let it be considered the sensible qualities are colour, figure, motion, smell, taste, etc., i. e., the ideas perceived by sense. Now, for an idea to exist in an unperceiving thing is a manifest contradiction; for to have an idea is all one as to perceive; that therefore wherein colour, figure, etc., exist must perceive them; hence it is clear there can be no unthinking substance or *substratum* of those ideas.

CHAPTER 41

BENJAMIN FRANKLIN

PROBABLY NO AMERICAN OF HUMBLE origin ever attained to more popularity and enduring fame than did Benjamin Franklin. He was born in Boston, Massachusetts, January 17, 1706, but Philadelphia, Pennsylvania claimed him at an early age. His formal education was finished in his eleventh year, when he began to work as a general utility boy in his father's shop. His fondness of reading and love of books led him into the printers' trade and although he was penniless when he arrived in Philadelphia, it was there that he was soon to become one of the foremost journalists of the times. At the age of 15 he was writing for the colonial press, contributing essays notable for their sensible moralizing and practical wisdom. His success at writing began with the publication of *Poor Richard's Almanac* which began in 1732, and his paper, *The Gazette*, became the most popular of colonial sheets. Later he wrote his *Autobiography*, which is considered one of the most important original contributions to American literature.

Franklin is described as the first great Man of Letters in America, and during the Revolutionary period was the foremost representative and diplomat of the colonists against the stamp duty and taxation without representation. He spent 40 years as a public official including serving as Postmaster General. He was one of the first signers of the Declaration of Independence, and was one of the framers of the Constitution. He was sent to France as a commissioner of the colonists and it was through his effort that aid was given the colonists by France.

As well as being a journalist, statesman, and diplomat, he was the first in America to organize a police force, city care of the streets, fire companies, a public library, and a city academy. He flew a kite in a thunder shower, drew down electricity, and invented the lightning rod. He was one of the eminent scientists of his day and was not only a member of the Royal Society, and one of the eight foreign members of the Royal Academy of Sciences in Paris, but was also a member of the Rosicrucian Order.

His eventful career ended April 17, 1790, great as a journalist and writer, as a scientist, a statesman, diplomat, and man of affairs.

Below we give an interesting excerpt of his writing on the subject of the relationship of lightning and electricity and his invention of the lightning rod.

THE IDENTITY OF LIGHTNING AND ELECTRICITY

We know that points have a property, by which they draw on as well as throw off the electrical fluid, at greater distances than blunt bodies can. That is, as the pointed part of an electrified body will discharge the atmosphere of the body, to communicate it farthest to another body, so the point of an unelectrified body will draw off the electrical atmosphere from an electrified body, farther than a blunter part of the same unelectrified body will do. Thus, a pin held by the head, and the point presented to an electrified body, will draw off its atmosphere at a foot distance; where, if the head were presented instead of the point, no such effect would follow. To understand this, we may consider, that, if a person standing on the floor would draw off the electrical atmosphere from an electrified body, an iron crow and a blunt knitting-needle, held alternately in his hand, and presented for that purpose, do not draw with different forces in proportion to their different masses. For the man, and what he holds in his hand, be it large or small, are connected with the common mass of unelectrified matter; and the force with which he draws is the same in both cases, it consisting in the different proportion of electricity in the electrified body, and that common mass. But the force, with which the electrified body retains its atmosphere by attracting it, is proportioned to the surface over which the particles are placed; that is, four square inches of that surface retain their atmosphere with four times the force that one square inch retains its atmosphere. And, as in plucking the hairs from the horse's tail, a degree of strength not sufficient to pull away a handful at once, could yet easily strip it hair by hair, so a blunt body presented cannot draw off a number of particles at once, but a pointed one, with no greater force, takes them away easily, particle by particle.

These explanations of the power and operation of points, when they first occurred to me, and while they first floated in my mind, appeared perfectly satisfactory; but now I have written them, and considered them more closely, I must own I have some doubts about them; yet, as I have at present nothing better to offer in their stead, I do not cross them out; for, even a bad solution read, and its faults discovered, has often given rise to a good one, in the mind of an ingenious reader.

Nor is it of much importance to us to know the manner in which nature executes her laws; it is enough if we know the laws themselves. It is of real use to know that China left in the air unsupported, will fall and break; but how it comes to fall, and why it breaks, are matters of speculation. It is a pleasure indeed to know them, but we can preserve our China without it.

Thus, in the present case, to know this power of points may possibly be of some use to mankind, though we should never be able to explain it. The following experiments, as well as those in my first paper, show this power. I have a large prime conductor, made of several thin sheets of clothier's pasteboard, formed into a tube, near ten feet long and a foot diameter. It is covered with Dutch embossed paper, almost totally gilt. This large metallic surface supports a much greater electrical atmosphere than a rod of iron of fifty times the weight would do. It is suspended by silk lines, and when charged will strike, at near two inches distance, a pretty hard stroke, so as to make one's knuckle ache. Let a person standing on the floor present the point of needle, at twelve or more inches distance from it, and while the needle is so presented, the conductor cannot be charged, the point drawing off the fire as fast as it is thrown on by the electrical globe. Let it be charged and then present the point at the same distance, and it will suddenly be discharged. In the dark you may see the light on the point, when the experiment is made. And if the person holding the point stands upon wax, he will be electrified by receiving the fire at that distance. Attempt to draw off the electricity with a blunt body, as a bolt of iron round at the end, and smooth, (a silversmith's iron punch, inch thick, is what I use,) and you must bring it within the distance of three inches before you can do it, and then it is done with a stroke and crack. As the pasteboard tube hangs loose on silk lines, when you approach it with the punch-iron, it likewise will move towards the punch, being attracted while it is charged; but if, at the same instant, a point be presented as before, it retires again, for the point discharges it. Take a pair of large brass scales, of two or more feet beam, the cords of the scales being silk. Suspend the beam by a packthread from the ceiling, so that the bottom of the scales may be about a foot from the floor; the scales will move round in a circle by untwisting of the packthread. Set the iron punch on the end upon the floor, in such a place as that the scales may pass over it in making their circle; then electrify one scale by applying the wire of a charged phial to it. As they move around, you see that scale draw nigher to the floor, and dip more when it comes over the punch; and, if that be placed at a proper distance, the scale will snap and discharge its fire into it. But, if a needle be stuck on the end of the punch, its point upward, the scale, instead of drawing nigh to the punch, and snapping, discharges its fire silently through the point, and rises higher from the punch. Nay, even if the needle be placed upon the floor near the punch, its point upwards, the end of the punch, though so much higher than the needle, will not attract the scale and receive its fire, for the needle will get it and convey it away, before it comes nigh enough for the punch to act. And this is constantly observable in these experiments, that the greater quantity of electricity on the point likewise will draw it off at a still greater distance.

Now if the fire of electricity and that of lightning be the same, as I have endeavored to show at large in a former paper, this pasteboard tube and these scales may represent electrified clouds. If a tube of only ten feet long will strike and discharge its fire on the punch at two or three inches distance, an electrified cloud of perhaps ten thousand acres may strike and discharge on the earth at a proportionately greater distance. The horizontal motion of the scales over the floor, may represent the motion of the clouds over the earth; and the erect iron punch, a hill or high building; and then we see how electrified clouds passing over hills or high buildings at too great a height to strike, may be attracted lower till within their striking distance. And, lastly, if a needle fixed on the punch with its point upright, or even on the floor below the punch, will draw the fire from the scale silently at a much greater than striking distance, and so prevent its descending towards the punch; or if in its course it would have come nigh enough to strike, yet being first deprived of its fire it cannot, and the punch is thereby secured from the stroke; I say, if these things are so, may not the knowledge of this power of points be of use to mankind, in preserving houses, churches, ships, etc., from the stroke of lightning, by directing us to fix, on the highest parts of those edifices, upright rods of iron made sharp as a needle, and gilt to prevent rusting, and from the foot of those rods a wire down the outside of the building into the ground or down round one of the shrouds of a ship, and down her side till it reaches the water? Would not these pointed rods probably draw the electrical fire silently out of a cloud before it came nigh enough to strike, and thereby secure us from that most sudden and terrible mischief?

To determine the question, whether the clouds that contain lightning are electrified or not, I would propose an experiment to be tried where it may be done conveniently. On the top of some high tower or steeple, place a kind of sentry-box, big enough to contain a man and an electrical stand. From the middle of the stand let an iron rod rise and pass out of the door, and then upright twenty or thirty feet, pointed very sharp at the end. If the electrical stand be kept dry and clean, a man standing on it, when such clouds are passing low, might be electrified and afford sparks, the rod drawing fire to him from a cloud. If any danger to the man should be apprehended (though I think there would be none), let him stand on the floor of his box, and now and then bring near to the rod the loop of a wire that has one end fastened to the leads, he holding it by a wax handle; so the sparks, if the rod is electrified, will strike from the rod to the wire, and not affect him.

CHAPTER 42

DAVID HUME

DAVID HUME, ONE OF THE modernists of the eighteenth century, was born in Edinburgh, Scotland, in April 1711. While he attacked and criticized much of the philosophy that was then popular, he was really not a destructionist, but a constructionist, although he often argued on the negative side of a subject. He started his career as a law student intending to follow that work, but the philosophical trend of his mind led him into speculative philosophy, and as early as 1739 he published a *Treatise of Human Nature*.

The philospher Berkeley was quite popular at the time among philosophical students, and Berkeley had argued that we might identify the self within us as being similar to the unknown something that was the cause of all sensations, and therefore, the inner self was an intellectuality. Hume attacked this, denied that there was any distinct self within, and tried to convey the idea that the unassociated and disconnected sensations and ideas of the human consciousness gave an impression of a self that was not real. The thought that there might be an inner intellectuality that was an entity, and it might be immortal, was also criticized by Hume with the argument that the same self within was merely a form of human consciousness in which thoughts were registered temporarily and was, therefore, the source of habit and of all mental phenomena.

About 1741 he became interested in political subjects, and in this field he gained great popularity, and his essays were widely read. Here again his philosophy was analytical as well as constructive, and he argued that the idea that money, rather than men and commodities, constituted wealth was an error of human judgment. He further argued against the idea that exports should exceed imports, and protested against the principle of restricted trade. Many of Hume's ideas along this line were later adopted by Adam Smith.

Hume was appointed librarian of the Faculty of Advocates in 1752, and then turned his attention toward history. Within the following ten years he compiled and issued his *History of England*.

In 1763 he became Secretary to Lord Hertford, Ambassador to France, and from 1767 to 1769 he was Under-Secretary of State. His transition occurred in the city of his birth on August 25, 1776.

HUME'S ARGUMENT AGAINST PERSONAL IDENTITY

There are some philosophers who imagine we are every moment intimately conscious of what we call our SELF; that we feel its existence and its continuance in existence; and are certain, beyond the evidence of a demonstration, both of its perfect identity and simplicity. The strongest sensation, the most violent passion, say they, instead of distracting us from this view, only fix it the more intensely, and make us consider their influence on self either by their pain or pleasure. To attempt a farther proof of this were to weaken its evidence; since no proof can be derived from any fact, of which we are so intimately conscious; nor is there any thing, of which we can be certain, if we doubt of this.

Unluckily all these positive assertions are contrary to that very experience, which is pleaded for them, nor have we any idea of self, after the manner it is here explained. For from what impression could this idea be derived? This question 'tis impossible to answer without a manifest contradiction and absurdity; and yet 'tis a question, which must necessarily be answered, if we would have the idea of self pass for clear and intelligible. It must be some one impression, that gives rise to every real idea. But self or person is not any one impression, but that to which our several impressions and ideas are supposed to have a reference. If any impression gives rise to the idea of self, that impression must continue invariably the same, through the whole course of our lives; since self is supposed to exist after that manner. But there is no impression constant and invariable. Pain and pleasure, grief and joy, passions and sensations succeed each other, and never all exist at the same time. It cannot, therefore, be from any of these impressions, or from any other, that the idea of self is derived; and consequently there is no such idea.

But farther, what must become of all our particular perceptions upon this hypothesis? All these are different, and distinguishable, and separable from each other, and may be separately considered and may exist separately, and have no need of any thing to support their existence. After what manner, therefore, do they belong to self; and how are they connected with it? For my part, when I enter most intimately into what I call *myself*, I always stumble on some particular perception or other, of heat or cold, light or shade, love or hatred, pain or pleasure. I never can catch *myself* at any time without a perception, and never can observe any thing but the perception. When my perceptions are removed for any time, as by sound sleep; so long am I insensible of myself, and may truly be said not to exist. And were all my perceptions removed by death, and could I neither think, nor feel, nor see, nor love, nor hate after the dissolution of my body. I should be entirely annihilated, nor do I conceive what is farther requisite to make me a perfect nonentity. If any one upon serious and unprejudiced reflection, thinks he has a different notion of himself, I must confess I can reason no longer with him. All I can allow him is, that he may be in the right as well as I, and that we are essentially different in this particular. He may, perhaps, perceive something simple and continued, which he calls himself, though I am certain there is no such principle in me.

But setting aside some metaphysicians of this kind, I may venture to affirm of the rest of mankind, that they are nothing but a bundle or collection of different perceptions, which succeed each other with an inconceivable rapidity, and are in a perpetual flux and movement. Our eyes cannot turn in their sockets without varying our perceptions. Our thought is still more variable than our sight; and all our other senses and faculties contribute to this change; nor is there any single power of the soul, which remains unalterably the same, perhaps for one moment. The mind is a kind of theater, where several perceptions successively make their appearance; pass, re-pass, glide away, and mingle in an infinite variety of postures and situations. There is properly no simplicity in it at one time, nor identity in different; whatever natural propension we may have to imagine that simplicity and identity. The comparison of the theater must not lead us. They are the successive perceptions only, that constitute the mind; nor have we the most distant notion of the place, where these scenes are represented, or of the materials, of which it is composed.

What then gives us so great a propension to ascribe an identity to these successive perceptions, and to suppose ourselves possest of an invariable and uninterrupted existence through the whole course of our lives? In order to answer this question, we must distinguish betwixt personal identity, as it regards our thought or imagination, and as it regards our passions or the concern we take in ourselves. The first is our present subject; and to explain it perfectly we must take the matter pretty deep, and account for that identity, which we attribute to plants and animals; there being a great analogy betwixt it, and the identity of a self or person.

We have a distinct idea of an object, that remains invariable and uninterrupted through a supposed variation of time; and this idea we call that of identity or sameness. We have also a distinct idea of several different objects existing in succession, and connected together by a close relation; and this to an accurate view affords as perfect a notion of diversity, as if there was no manner of relation among the objects. But though these two ideas of identity, and a succession of related objects be in themselves perfectly distinct, and even contrary, yet 'tis certain, that in our common way of thinking they are generally confounded with each other. That action of the imagination, by which we consider the uninterrupted and invariable object, and that by which we reflect on the succession of related objects, are almost the same to the feeling, nor is there much more effort of thought required in the latter case than in the former. The relation facilitates the transition of the mind from one object to another and renders its passage as smooth as if it contemplated one continued object. This resemblance is the cause of the confusion and mistake, and makes us substitute the notion of identity, instead of that of related objects. However at one instant we may consider the related succession as variable or interrupted, we are sure the next to ascribe to it a perfect identity, and regard it as invariable and uninterrupted. Our propensity to this mistake is so great from the resemblance above-mentioned, that we fall into it before we are aware; and though we incessantly correct ourselves by reflection, and return to a more accurate method of thinking, yet we cannot long sustain our philosophy, or take off this bias from the imagination. Our last resource is to yield to it, and boldly assert that these different related objects are in effect the same, however interrupted and variable. In order to justify to ourselves this absurdity, we often feign some new and unintelligible principle, that connects the objects together, and prevents their interruption or variation. Thus we feign the continued existence of the perceptions of our senses, to remove the interruption; and run into the notion of a soul, and self, and substance, to disguise the variation. But we may farther observe, that where we do not give rise to such a fiction, our propension to confound identity with relation is so great, that we are apt to imagine something unknown and mysterious, connecting the parts, beside their relation; and this I take to be the case with regard to the identity we ascribe to plants and vegetables. And even when this does not take place, we still feel a propensity to confound these ideas, though we are not able fully to satisfy ourselves in that particular, nor find anything invariable and uninterrupted to justify our notion of identity.

CHAPTER 43

JOHANN GOTTFRIED VON HERDER

JOHANN GOTTFRIED VON HERDER was one of the outstanding German classical writers of his period. He was imbued with the ideal of reforming the philosophy and theology of his period. He was of a highly sensitive nature and worked with great ardor on every new subject to which he turned his attention, but unfortunately, his zeal did not endure and many of his great works are incomplete. He left a definite impression upon the peoples of his time, and conveyed some profound thoughts which are more appreciated now than during his own time. He was born August 25, 1744, at Mohrungen, East Prussia. He was originally schooled and prepared to be a surgeon, but upon witnessing his first operation, fainted and turned thereafter to theology. In 1764 he was appointed teacher and preacher in a Cathedral school at Riga. In 1771 the Grand Duke appointed him Court Preacher and Counselor of the upper consistory at Bückeburg, and in 1776, through Goethe's influence, he obtained posts at Weimar. He was a great admirer of Kant's writings, yet seemed extremely critical of them at times. Although of a mild disposition, he was quite vicious in his attacks on his enemies. Students of metaphysics will all enjoy reading his works, for they will find in them a keen sense of the mystical. Below is an excerpt from one of his prominent writings entitled *Man a Link Between Two Worlds.*

MAN A LINK BETWEEN TWO WORLDS

Everything in Nature is connected: one state pushes forward and prepares another. If, then, man be the last and highest link, closing the chain of terrestrial organization, he must begin the chain of a higher order of creatures as its lowest link, and is probably, therefore, the middle ring between the two adjoining systems of the creation. He cannot pass into any other organization upon earth without turning backward and wandering in a circle. That he should stand still is impossible; since no living power in the dominions of the most active goodness is at rest; thus there must be a step before him, close to him, yet as exalted above him as he is preeminent over the brute, to whom he is at the same time nearly allied.

This view of things, which is supported by all the laws of nature, alone gives us the key to the wonderful phenomenon of man, and at the same time to the only philosophy of his history. . . .

Far as the life of man here below is from being calculated for entirety; equally far is this incessantly revolving sphere from being a repository of permanent works of art, a garden of never-fading plants, a seat to be eternally inhabited. We come and go: every moment brings thousands into the world, and takes thousands out of it. The Earth is an inn for travellers; a planet, on which birds of passage rest themselves, and from which they hasten away. The brute lives out his life; and if his years be too few to attain higher ends, his inmost purpose is accomplished: his capacities exist, and he is what he was intended to be. Man alone is in contradiction with himself, and with the Earth: for, being the most perfect of all creatures, his capacities are the farthest from being perfected, even when he attains the longest term of life before he quits the world. But the reason is evident: his state, being the last upon this Earth, is the first in another sphere of existence, with respect to which he appears here as a child making his first essays. Thus he is the representative of two worlds at once; and hence the apparent duplicity of his essence. . . .

If superior creatures look down on us, they may view us in the same light as we do the *middle species*, with which Nature makes a transition from one element to another. The ostrich flaps his feeble wings to assist himself in running, but they cannot enable him to fly; his heavy body confines him to the ground. Yet the organizing Parent has taken care of him, as well as of every middle creature; for they are all perfect in themselves, and only appear defective to our eyes. It is the same with man here below: his defects are perplexing to an earthly mind; but a superior spirit that inspects the internal structure, and sees more links of the chain, may indeed pity, but cannot despise him. He perceives why man must quit the world in so many different states, young and old, wise and foolish, grown gray in second childhood, or an embryo yet unborn. Omnipotent goodness embraces madness and deformity, and all the degrees of cultivation, and all the errors of man, and wants not balsams to heal the wounds that death alone could mitigate. Since probably the future state springs out of the present, as our organization from inferior ones, its business is no doubt more closely connected with our existence here than we imagine. The garden above blooms only with plants of which the seeds have been sown here, and put forth their first germs from coarser husk. If, then, as we have seen, sociality, friendship, or active participation in the pains and pleasures of others, be the principal end to which humanity is directed, the finest flower of human life must necessarily there attain the vivifying form, the overshadowing height, for which our heart thirsts in vain in any earthly situation. Our bretheren above, therefore, assuredly love us with more warmth and purity of affection than we can bear to them: for they see our state more clearly; to them the moment of time is no more, all discrepancies are harmonized, and in us they are probably educating unseen partners of their happiness, and companions of their labors. But one step farther, and the oppressed spirit can breathe more freely, the wounded heart recovers: they see the passenger approach it, and stay his sliding feet with a powerful hand.

Since, therefore, we are of a middle species between two orders, and in some measure partake of both, I cannot conceive that the Future state is so remote from the Present, and so incommunicable with it, as the animal part of man is inclined to suppose, and indeed many steps and events in the history of the human race are to me incomprehensible, without the operation of superior influence. A divine economy has certainly ruled over the human species from its first origin, and conducted him into the course the readiest way....

This much is certain, that there dwells an infinity in each of man's powers, which cannot be developed here, where it is repressed by other powers, by animal senses and appetites, and lies bound as it were to the state of terrestrial life. Particular instances of memory, of imagination, nay, of prophesy and prehension, have discovered wonders of that hidden treasure which reposes in the human soul; and indeed the senses are not to be excluded from this observation. That diseases and partial defects, have been the principal occasions of indicating this treasure alters not the nature of the case; since this very disproportion was requisite to set one of the weights at liberty, and display its power.

The expression of Leibnitz, that the soul is a mirror of the universe, contains perhaps a more profound truth than has usually been educed from it: for the powers of a universe seem to lie concealed in her, and require only an organization, or a series of organizations, to set them in action. Supreme goodness will not refuse her this organization, but guides her like a child in leading strings, gradually to prepare her for the fulness of increasing enjoyment, under a persuasion that her powers and senses are self-acquired. Even in her present fetters *space* and *time* are to her empty words: they measure and express relations of the body, but not of her internal capacity, which extends beyond time and space, when it acts in perfect internal quiet. Give thyself no concern for the place and hour of thy future existence: the Sun, that enlightens thy days, is necessary to thee during thy abode and occupation upon earth; and so long it obscures all the celestial stars. When it sets, the universe will appear in greater magnitude; the sacred night, that once enveloped thee, and in which thou wilt be enveloped again, covers thy Earth with shade, and will open to thee the splendid volume of immortality in Heaven. There are habitations, worlds and spaces, that bloom in unfading youth, though ages on ages have rolled over them, and defy the changes of time and season; but everything that appears to our eyes decays, and perishes, and passes away; and all the pride and happiness of Earth are exposed to inevitable destruction.

This earth will be no more, when thou thyself still art, and enjoyest God and His creation in other abodes, and differently organized. On it thou hast enjoyed much good. On it thou hast attained an organization, in which thou hast learned to look around and above thee as a child of Heaven. Endeavor, therefore, to leave it contentedly, and bless it in the field, where thou hast sported as a child of immortality, and as the school, where thou hast been brought up in joy, and in sorrow, to manhood. Thou hast no farther claim on it; it has no farther claim on thee. As the flower stands erect, and closes the realm of the subterranean inanimate creation, to enjoy the commencement of life, in the region of day; so is man raised above all the creatures that are bowed down to the Earth. With uplifted eye, and outstretched hand, he stands as a son of the family, awaiting his father's call.

CHAPTER 44

PIERRE SIMON LAPLACE

PIERRE SIMON LAPLACE WAS born in Normandy, March 28, 1749. Even as a youth of eighteen years did he show signs of brilliance. At that young age he became teacher of mathematics at Beaumont. From 1770, for a number of years, he was extremely busy with Lagrange in establishing the stability of the solar system, accounting for its perturbations and interactions, and attempting to prove that they were periodic and not adventitious.

His work entitled *Mechanics of the Heavens* was a gigantic exposition of the movements of the solar system, but the work which made him famous and for which all astronomers pay him a debt of honor is his *System of the World.* In this work he advances his famous nebular hypothesis, in which he accounts for the origin of the universe. It attracted considerable attention at the time because of its logical explanation of cosmological mysteries, and because it was the most sound cosmological theory up to the time. His previous work had consisted of tracing the law of gravitation throughout the system of planets.

Laplace also tried being a politician, but he failed miserably. Politically he changed as often as the wind. During Napoleon's regime his views made him a count. On the other hand, during the regime of the restored Bourbon king in 1817 he was made a marquis. He died March 5, 1827.

NEBULAR HYPOTHESIS

In the primitive state in which we have supposed the Sun to be, it resembles those substances which are termed nebulae, which, when seen through telescopes, appear to be composed of a nucleus, more or less brilliant, surrounded by a nebulosity, which, by condensing on its surface, transforms it into a star. If all the stars are conceived to be similarly formed, we can suppose their anterior state of nebulosity to be preceded by other states, in which the nebulous matter was more or less diffuse, the nucleus being at the same time more or less brilliant. By going back in this manner, we shall arrive at a state of nebulosity so diffuse, that its existence can with difficulty be conceived.

For a considerable time back, the particular arrangement of some stars visible to the naked eye, has engaged the attention of philosophers. Mitchel remarked long since how extremely improbable it was that the stars composing the constellation called the Pleiades, for example, should be confined within the narrow space which contains them, by the sole chance of hazard; from which he inferred that this group of stars, and the similar groups which the heavens present to us, are the effects of a primitive law of nature. These groups are a general result of the condensation of nebulae of several nuclei; for it is evident that the nebulous matter being perpetually attracted by these different nuclei, ought at length to form a group of stars, like to that of the Pleiades. The condensation of nebulae consisting of two nuclei, will in like manner form stars very near to each other, revolving the one about the other like to the double stars, whose respective motions have been already recognized.

But in what manner has the solar atmosphere determined the motions of rotation and revolution of the planets and satellites? If these bodies had penetrated deeply into this atmosphere, its resistance would cause them to fall on the Sun. We may therefore suppose that the planets were formed at its successive limits, by the condensation of zones of vapours, which it must, while it was cooling, have abandoned in the plane of its equator.

Let us resume the results which we have given in the tenth chapter of the preceding book. The Sun's atmosphere cannot extend indefinitely; its limit is the point where the centrifugal force arising from the motion of rotation balances the gravity; but according as the cooling contracts the atmosphere, and condenses the molecules which are near to it, on the surface of the star, the motion of rotation increases; for, in virtue of the principle of areas, the sum of the areas described by the *radius vector* of each particle of the Sun and its atmosphere, and projected on the plane of its equator, is always the same. Consequently the rotation ought to be quicker, when these particles approach to the centre of the Sun. The centrifugal force arising from this motion becoming thus greater; the point where the gravity is equal to it, is nearer to the centre of the Sun. Supposing, therefore, what is natural to admit, that the atmosphere extended at any epoch as far as this limit, it ought, according as it cooled, to abandon the molecules, which are situated at this limit, and at the successive limits produced by the increased rotation of the Sun. These particles, after being abandoned, have continued to circulate about this star, because their centrifugal force was balanced by their gravity. But as this equality does not obtain for these molecules of the atmosphere which are situated on the parallels to the Sun's equator, these have come nearer by their gravity to the atmosphere according as it condensed, and they have not ceased to belong to it inasmuch as by their motion, they have approached to the plane of this equator.

Let us now consider the zones of vapours, which have been successively abandoned. These zones ought, according to all probability, to form by their condensation, and by the mutual attraction of their particles, several concentrical rings of vapours circulating about the Sun. The mutual friction of the molecules of each ring ought to accelerate some and retard others, until they all had acquired the same angular motion. Consequently the real velocities of the molecules which are farther from the Sun, ought to be greatest. The following cause ought likewise to contribute to this difference of velocities: The most distant particles of the Sun, and which, by the effects of cooling and condensation, have collected so as to constitute the superior part of the ring, have always described areas proportional to the times, because the central force by which they are actuated has been constantly directed to this star; but this constancy of areas requires an increase of velocity, according as they approach more to each other. It appears that the same cause ought to diminish the velocity of the particles, which, situated near the ring, constitute its inferior part.

If all the particles of a ring of vapours continued to condense without separating, they would at length constitute a solid or a liquid ring. But the regularity which this formation requires in all the parts of the ring, and in their cooling, ought to make this phenomenon very rare. Thus the solar system presents but one example of it; that of the rings of Saturn. Almost always each ring of vapours ought to be divided into several masses, which, being moved with velocities which differ little from each other, should continue to revolve at the same distance about the Sun. These masses should assume a spheroidical form, with a rotatory motion in the direction of that of their revolution, because their inferior particles have a less real velocity than the superior; they have therefore constituted so many planets in a state of vapour. But if one of them was sufficiently powerful, to unite successively by its attraction, all the others about its centre, the ring of vapours would be changed into one sole spheroidical mass, circulating about the Sun, with a motion of rotation in the same direction with that of revolution. This last case has been the most common; however, the solar system presents to us the first case, in the four small planets which revolve between Mars and Jupiter, at least unless we suppose with Olbers, that they originally formed one planet only, which was divided by an explosion into several parts, and actuated by different velocities. Now if we trace the changes which a further cooling ought to produce in the planets formed of vapours, and of which we have suggested the formation, we shall see to arise in the centre of each of them, a nucleus increasing continually, by the condensation of the atmosphere which environs it. In this state, the planet resembles the Sun in the nebulous state, in which we have first supposed it to be; the cooling should therefore produce at the different limits of its atmosphere, phenomena similar to those which have been described, namely, rings and satellites circulating about its centre in the direction of its motion of rotation, and revolving in the same direction on their axes. The regular distribution of the mass of rings of Saturn about its centre and in the plane of its equator, results naturally from this hypothesis, and, without it, is inexplicable. Those rings appear to me to be existing proofs of the primitive extension of the atmosphere of Saturn, and of its successive condensations. Thus, the singular phenomena of the small eccentricities of the orbits of the planets and satellites, of the small inclination of these orbits to the solar equator, and of the identity in the direction of the motions of rotation and revolution of all those bodies with that of the rotation of the Sun, follow the hypothesis which has been suggested, and render it extremely probable. If the solar system was formed with perfect regularity, the orbits of the bodies which compose it would be circles, of which the planes, as well as those of the various equators and rings, would coincide with the plane of the solar equator. But we may suppose that the innumerable varieties which must necessarily exist in the temperature and density of different parts of these great masses, ought to produce the eccentricities of their orbits, and the deviations of their motions, from the plane of this equator.

In the preceding hypothesis, the comets do not belong to the solar system. If they be considered, as we have done, as small nebulae, wandering from one solar system to another, and formed by the condensation of the nebulous matter, which is diffused so profusely throughout the universe, we may conceive that when they arrive in that part of space where the attraction of the Sun predominates, it should force them to describe elliptic or hyperbolic orbits. But as their velocities are equally possible in every direction, they must move indifferently in all directions, and at every possible inclination to the elliptic; which is conformable to observation. Thus the condensation of the nebulous matter, which explains the motions of rotation and revolution of the planets and satellites in the same direction, and in orbits very little inclined to each other, likewise explains why the motions of the comets deviate from this general law.

CHAPTER 45

JOHANN GOTTLIEB FICHTE

JOHANN GOTTLIEB FICHTE WAS born May 19, 1962, in Upper Lusatia, Germany. He was educated in the lower common schools under the patronage of the Baron Von Miltitz. Later, from 1780 to 1787, at Jena and Leipzig, he studied theology, during which time he supported himself by tutoring. In 1792 he published the work, *The Critique of All Revelation*, which caused quite a sensation. It was attributed to Kant as, due to an error, Fichte's name was omitted. Kant corrected the mistake, admitting it was not his own, and at the same time, praised Fichte. The approbation by Kant resulted in the making of Fichte.

Fichte later took an active interest in politics, and wrote in an attempt to justify the French Revolution. In 1794 he completed his *Science of Knowledge* in which he demonstrated Kant's system of conception by an analysis of consciousness. Considerable criticism was heaped upon him because of his statement that a moral order or system was the equivalent of the idea of God. He eventually fled for refuge to Berlin where, with the aid of his students, he incited the Germans in an uprising in 1808 against Napoleon. He died of a hospital fever on January 27, 1814.

The essence of his philosophy is: Though being or phenomena may exist apart from man, it receives its identity or its form from the human consciousness; and the soul of man is merely man's consciousness of the conduct and acts of his ego, or, in other words, the realization of the life force within himself.

OUTLINES OF THE DOCTRINE OF KNOWLEDGE

I

The Doctrine of Knowledge, apart from all special and definite knowing, proceeds immediately upon Knowledge itself, in the essential unity in which it recognizes Knowledge as existing; and it raises this question in the first place: How this Knowledge can come into being, and what it is in its inward and essential Nature?

The following must be apparent: There is but One who exists absolutely by and through himself,—namely, God; and God is not the mere dead conception to which we have thus given utterance, but he is in Himself pure life. He can neither change nor determine himself in aught within himself, nor become any other Being; for his Being contains within it all his Being and all possible Being, and neither within him nor out of him can any new Being arise.

If, therefore, Knowledge must be, and yet be not God Himself, then, since there is nothing but God, it can only be God out of himself,—God's Being out of his Being,—his Manifestation, in which he dwells wholly as he is in himself, while within himself he also still remains wholly as he is. But such a Manifestation is a picture or *Schema*.

If there be such a Manifestation—and this can only become evident through its immediate being, seeing that it is immediate—it can only be because God is: and, so surely as God is, it cannot but be. It is, however, by no means to be conceived of as a work of God, effected by some particular act, whereby a change is wrought in himself; but it is to be conceived of as an immediate consequence of his Being. It is absolutely, according to the Form of his Being, just as he himself is absolutely; although it is not he himself, but his Manifestation.

Again:—Out of God there can be nothing whatever but this;—no Being that is essentially independent, for that he alone is;—only his Manifestation can there be out of him, and thus a Being out of God signifies merely his Manifestation;—the two expressions mean precisely the same thing.

II

Further:—Since it cannot be overlooked by the Doctrine of Knowledge that Actual Knowledge does by no means present itself as a Unity, such as is assumed above, but as a Multiplicity, there is consequently a second task imposed upon it,—that of setting forth the ground of this apparent Multiplicity. It is of course understood that this ground is not to be derived from any outward source, but must be shown to be contained in the essential Nature of Knowledge itself as such;—and that therefore this problem, although apparently twofold, is yet but one and the same,—namely, to set forth the essential Nature of Knowledge.

III

This Being out of God cannot, by any means, be a limited, completed, and inert Being, since God himself is not such a dead Being, but, on the contrary, is Life;—but it can only be a Power, since only a Power is the true formal picture or Manifestation of Life. And indeed it can only be the Power of realizing that which is contained in itself—a Manifestation. Since this Power is the expression of a determinate Being—the Manifestation of the Divine Life—it is itself determined; but only in the way in which an absolute Power may be determined,—by laws, and indeed by determinate laws. If this or that is to become actual, the Power must operate in this way or that, subject to that determination.

IV

This in the first place:—There can be an actual Being out of God only through the self-realization of this absolute Power:—this Power, however, can only produce pictures or Manifestations, which by combination become Actual Knowledge. Thus, whatever exists out of God, exists only by means of absolutely free Power, as the Knowledge belonging to this Power, and in its Knowledge;—and any other Being but this out of the true Being which lies hidden in God is altogether impossible.

V

Again, as to the determination of this Power by laws:—It is, in the first place, determined through itself, as the Power of Actual Knowledge. But it is essential to Actual Knowledge that some particular Manifestation should be realized through this Power; and then that through the same identical Power, in the same identical position, this Manifestation should be recognized as a Manifestation, and as a Manifestation not in itself independent, but demanding, as a condition of its existence, a Being out of itself. The immediate and concrete expression of this recognition,—which in Actual Knowledge never attains to consciousness, but which is elevated into consciousness only by means of the Doctrine of Knowledge,—is Actual Knowledge itself in its Form; and, in consequence of this latter recognition, there is, of necessity, assumed an Objective Reality, wholly transcending the Manifestation and independent of Knowledge. Since in this Knowledge of the Objective Reality, even the Manifestation itself is concealed, much more is the Power which creates it concealed and unseen. This is the fundamental law of the Form of Knowledge. So surely therefore as the Power develops itself in this particular way, it develops itself as we have described; not merely Manifesting, but also manifesting the Manifestation as a Manifestation, and recognizing it in its dependent nature;—not that it must unconditionally do this, but that only by means of this process can it attain to Actual Knowledge.

In consequence of this there is much that remains invisible in Actual Knowledge, but which, nevertheless, really is as the manifestation of this Power. If therefore this, and all other manifestation of this Power, were to be imported into Knowledge, then could this only occur in a Knowledge other than that first mentioned; and thus would the unity of knowledge necessarily be broken up into separate parts, by the opposition of the law of the form of visibility to that law by which Knowledge perceives itself as a perfect and indivisible whole.

VI

Further:—Within this its Formal Being, this Power is also determined by an unconditional Imperative. It shall recognize itself as the Manifestation of the Divine Life, which it is originally, and through which alone it has Existence;—consequently this is its absolute vocation, in which its efficiency as a Power is completely exhausted. It shall recognize itself as the Manifestation of the Divine Life,—but it is originally nothing more than a Power, although most assuredly it is this determinate Power of the Manifestation of God:—if it is to recognize itself as such a Manifestation in Reality, then it must make itself so actually by the realization of the Power—by its self-realization.

VII

The recognition of itself as a Power to which an unconditional Imperative is addressed, and which is able to fulfill that Imperative, and the actual realization of this Power, should the latter come to pass, are distinct from each other; and the possibility of the latter is dependent on the previous accomplishment of the former.

It shall recognize itself as the Divine Manifestation, not by means of any Being inherent in itself, for there is no such Being, but by means of the realization of the Power. It must therefore previously possess the knowledge that it is such a Power, and also by what marks it may recognize itself in its self-realization, in order that it may direct its attention to these characteristic marks, and so be enabled to judge of the realization which they denote.

Or it may be regarded thus:—By means of the realization of the Power there arises a Manifestation, and a consciousness of that which is contained in the Manifestation, and not more than this. The formal addition, which lies beyond the immediate contents of the Manifestation,—i.e., that it is the Manifestation of God,—is not immediately contained in it; and can only be attributed to it in consequence of some characteristic mark perceived in the actual realization of the Power. The characteristic mark is this:—that the Power realize itself, with absolute Freedom, in accordance with the recognized universal Imperative.

CHAPTER 46

SIR HUMPHRY DAVY

SIR HUMPHRY DAVY ATTAINED prominence as a chemist and had, as well, a deep interest in poetry and metaphysics—spheres of human expression usually considered quite opposite to each other. He was born at Penzance, Cornwall, England, December 17, 1778. He began the study of medicine in 1795, but was won over to chemistry. He later became lecturer at the Royal Institution, then newly established. He was both brilliant and forceful. He gained prominence by the invention of what is known as the "safety lamp" for miners, preventing the previously common explosions in mines from what was termed by the miners as "fire-damp" or ignitible gases. He was knighted for distinguished service to his country in 1812, and became a Baronet in 1818. He was president of the Royal Society, an organization whose early years figure in Rosicrucian history. It has been said of him that if he had not won recognition as an outstanding chemist, he could have as a poet and metaphysician, because of his poetic temperament. Examples of his mystical insight are given below in the brief essays. Every lover of mysticism and metaphysics will enjoy his cogent statements.

THE OFFICE OF PAIN

The laws of nature are all directed by Divine Wisdom for the purpose of preserving life, and increasing happiness. Pain seems in all cases to precede the mutilation or destruction of those organs which are essential to vitality, and for the end of preserving them; but the mere process of dying seems to be the falling into a deep slumber; and in animals, who have no fear of death dependent upon imagination, it can hardly be accompanied by very intense suffering. In the human being, moral and intellectual motives constantly operate in enhancing the fear of death, which, without these motives in a reasoning being, would probably become null, and the love of life be lost upon every slight occasion of pain or disgust. But imagination is creative with respect to both these passions, which, if they exist in animals, exist independent of reason, or as instincts.

Pain seems intended by an all-wise Providence to prevent the *dissolution* of organs, and cannot follow their *destruction*. I know several instances in which the process of death has been observed, even to its termination by good philosophers; and the instances are worth repeating: Dr. Cullen, when dying, is said to have faintly articulated to one of his intimates, "I wish I had the power of writing or speaking; for then I would describe to you how pleasant a thing it is to die." Dr. Black—worn out by age, and a disposition to pulmonary hemorrhage, which obliged him to live very low—whilst eating his customary meal of bread and milk, fell asleep, and died in so tranquil a manner that he had not even spilt the contents of the cup which rested on his knee. And the late Sir Charles Blagden, whilst at a social meal, with his friends, Monsieur and Madame Bertholt and Gay Lussac, died in his chair so quietly, that not a drop of the coffee in the cup which he held in his hand, was spilt.

Indestructibility of Mind

The doctrine of the materialists was always, even in my youth, a cold, heavy, dull, and insupportable doctrine to me, and necessarily tending to atheism. When I had heard with disgust, in the dissecting-rooms, the plan of the physiologist, of the gradual accretion of matter, and its becoming endowed with irritability, ripening into sensibility, and acquiring such organs as were necessary by its own inherent forces, and at last issuing into intellectual existence, a walk into the green fields or woods, by the banks of rivers, brought back my feelings from nature to God. I saw in all the powers of matter the instruments of the Deity. The sunbeams, the breath of the zephyr, awakening animation in forms prepared by divine intelligence to receive it, the insensate seed, the slumbering eggs which were to be vivified, appeared, like the new-born animal, works of a divine mind; I saw love as the creative principle in the material world, and this love only as a divine attribute. Then my own mind I felt connected with new sensations and indefinite hopes—a thirst for immortality; the great names of other ages and of distant nations appeared to me to be still living around me, and even in the fancied movements of the heroic and the great, I saw, as it were, the degrees of the indestructibility of mind. These feelings, though generally considered as poetical, yet, I think, offer a sound philosophical argument in favor of the immortality of the soul. In all the habits and instincts of young animals, their feelings and movements, may be traced an intimate relation to their improved perfect state; their sports have always affinities to their modes of hunting or catching their food; and young birds even in the nests, show marks of fondness which, when their frames are developed, become signs of actions necessary to the reproduction and preservation of the species. The desire of glory, of honor, of immortal fame, and of constant knowledge, so usual in young persons of well-constituted minds, cannot, I think, be other than symptoms of the infinite and progressive nature of the intellect—hopes which, as they cannot be gratified here, belong to a frame of mind suited to a nobler state of existence.

Religion, whether natural or revealed, has always the same beneficial influence on the mind. In youth, in health and prosperity, it awakens feelings of gratitude and sublime love, and purifies at the same time that it exalts. But it is in misfortune, in sickness, in age, that its effects are most truly and beneficially felt; when submission in faith and humble trust in the divine will, from duties become pleasures, undecaying sources of consolation. Then it creates powers which were believed to be extinct; and gives a freshness to the mind, which was supposed to have passed away forever, but which is now renovated as an immortal hope. Then it is the Pharos, guiding the wave-tossed mariner to his home—as the calm and beautiful still basins of fiords, surrounded by tranquil groves and pastoral meadows, to the Norwegian pilot escaping from a heavy storm in the North Sea—or as the green and dewy spot, gushing with fountains, to the exhausted and thirsty traveller in the midst of the desert. Its influence outlives all earthly enjoyments, and becomes stronger as the organs decay and the frame dissolves. It appears as that evening-star of light in the horizon of life, which, we are sure, is to become, in another season, a morning-star; and it throws its radiance, through the gloom and shadow of death.

Intimations of a Future Life

Music is the sensual pleasure which approaches nearest to an intellectual one, and may probably represent the delight resulting from the perception of the harmony of things, and of truth as seen in God. The palm as an evergreen tree, and the amaranth as a perdurable flower, are emblems of immortality. If I am allowed to give a metaphorical allusion to the future state of the blest, I should imagine it by the orange grove in that sheltered glen, on which the sun is now beginning to shine, and of which the trees are at the same time loaded with sweet golden fruit and balmy silver flowers. Such objects may well portray a state in which hope and fruition become one eternal feeling.

CHAPTER 47

ARTHUR SCHOPENHAUER

ARTHUR SCHOPENHAUER WAS BORN in Danzig, Germany, February 22, 1788. His father was a prominent banker and highly respected in the district in which he lived. His mother was a writer of novels and books of travel. Young Schopenhauer entered the University of Göttingen in 1809. There he studied philosophy under the capable tutorship of Schulze, the Skeptic.

He was an ardent student of the writings of Kant and Plato. In 1811 he attended a series of lectures in Berlin by Fichte. He preferred the natural sciences.

Schopenhauer's graduation dissertation was "On the Fourfold Root of the Principle of Sufficient Reason." In 1814, at the age of 26, he had become quite pessimistic and disputatious, and he is known as the pessimistic philosopher. In 1819 he published his magnum opus entitled, *The World as Will and Idea.*

He died September 21, 1860, just about the time his works were becoming extremely popular.

ON MEN OF LEARNING

When one sees the number and variety of institutions which exist for the purpose of education, and the vast throng of scholars and masters, one might fancy the human race to be very much concerned about truth and wisdom. But here, too, appearances are deceptive. The masters teach in order to gain money, and strive, not after wisdom, but the outward show and reputation of it; and the scholars learn, not for the sake of knowledge and insight, but to be able to chatter and give themselves airs. Every thirty years a new race comes into the world—a youngster that knows nothing about anything, and after summarily devouring in all haste the results of human knowledge as they have been accumulated for thousands of years, aspires to be thought cleverer than the whole of the past. For this purpose he goes to the University, and takes to reading books—new books, as being of his own age and standing. Everything he reads must be briefly put, must be new! he is new himself. Then he falls to and criticises. And here I am not taking the slightest account of studies pursued for the sole object of making a living.

Students, and learned persons of all sorts and every age, aim as a rule at acquiring information rather than insight. They pique themselves upon knowing about everything—stones, plants, battles, experiments, and all the books in existence. It never occurs to them that information is only a means of insight, and in itself of little or no value; that it is his way of thinking that makes a man a philosopher. When I hear of these portents of learning and their imposing erudition, I sometimes say to myself: Ah, how little they must have had to think about, to have been able to read so much! And when I actually find it reported of the elder Pliny that he was continually reading or being read to, at table, on a journey, or in his bath, the question forces itself upon my mind, whether the man was so very lacking in thought of his own that he had to have alien thought incessantly instilled into him; as though he were a consumptive patient taking jellies to keep himself alive. And neither his undiscerning credulity nor his inexpressibly repulsive and barely intelligible style—which seems like of a man taking notes, and very economical of paper—is of a kind to give me a high opinion of his power of independent thought.

We have seen that much reading and learning is prejudicial to thinking for oneself; and, in the same way, through much writing and teaching, a man loses the habit of being quite clear, and therefore thorough, in regard to the things he knows and understands; simply because he has left himself no time to acquire clearness or thoroughness. And so, when clear knowledge fails him in his utterances, he is forced to fill out the gaps with words and phrases. It is this, and not the dryness of the subject-matter, that makes most books such tedious reading. There is a saying that a good cook can make a palatable dish even out of an old shoe; and a good writer can make the dryest things interesting.

With by far the largest number of learned men, knowledge is a means, not an end. That is why they will never achieve any great work; because, to do that, he who pursues knowledge must pursue it as an end, and treat everything else, even existence itself, as only a means. For everything which a man fails to pursue for its own sake is but half-pursued; and true excellence, no matter in what sphere, can be attained only where the work has been produced for its own sake alone, and not as a means to further ends.

And so, too, no one will ever succeed in doing anything really great and original in the way of thought, who does not seek to acquire knowledge for himself, and making this the immediate object of his studies, decline to trouble himself about the knowledge of others. But the average man of learning studies for the purpose of being able to teach and write. His head is like a stomach and intestines which let the food pass through them undigested. That is just why his teaching and writing is of so little use. For it is not upon undigested refuse that people can be nourished, but solely upon the milk which secretes from the very blood itself.

The wig is the appropriate symbol of the man of learning, pure and simple. It adorns the head with a copious quantity of false hair, in lack of one's own; just as erudition means endowing it with a great mass of alien thought. This, to be sure, does not clothe the head so well and naturally, nor is it so generally useful, nor so suited for all purposes, nor so firmly rooted; nor when alien thought is used up, can it be immediately replaced by more from the same source, as is the case with that which springs from soil of one's own. So we find Sterne, in his *Tristram Shandy,* boldly asserting that *an ounce of a man's own wit is worth a ton of other people's.*

And in fact the most profound erudition is no more akin to genius than a collection of dried plants is like Nature, with its constant flow of new life, ever fresh, ever young, ever changing. There are no two things more opposed than the childish naiveté of an ancient author and the learning of his commentator.

Dilettanti, dilettanti! This is the slighting way in which those who pursue any branch of art or learning for the love and enjoyment of the thing,—*per il loro diletto,* are spoken of by those who have taken it up for the sake of gain, attracted solely by the prospect of money. This contempt of theirs comes from the base belief that no man will seriously devote himself to a subject, unless he is spurred on to it by want, hunger, or else some form of greed. The public is of the same way of thinking; and hence its general respect for professionals and its distrust of *dilettanti.* But the truth is that the *dilettante* treats his subject as an end, whereas the professional, pure and simple, treats it merely as a means. He alone will be really in earnest about a matter, who has a direct interest therein, takes to it because he likes it, and pursues it *con amore.* It is these, and not hirelings, that have always done the greatest work.

In the republic of letters it is as in other republics; favor is shown to the plain man—he who goes his way in silence and does not set up to be cleverer than others. But the abnormal man is looked upon as threatening danger; people band together against him, and have, oh! such a majority on their side.

The condition of this republic is much like that of a small State in America, where every man is intent only upon his own advantage, and seeks reputation and power for himself, quite heedless of the general weal, which then goes to ruin. So it is in the republic of letters; it is himself, and himself alone, that a man puts forward, because he wants to gain fame. The only thing in which all agree is in trying to keep down a really eminent man, if he should chance to show himself, as one who would be a common peril. From this it is easy to see how it fares with knowledge as a whole.

Between professors and independent men of learning there has always been from of old a certain antagonism, which may perhaps be likened to that existing between dogs and wolves. In virtue of their position, professors enjoy great facilities for becoming known to their contemporaries. Contrarily, independent men of learning enjoy, by their position, great facilities for becoming known to posterity; to which it is necessary that, amongst other and much rarer gifts, a man should have a certain leisure and freedom. As mankind takes a long time in finding out on whom to bestow its attention, they may both work together side by side.

He who holds a professorship may be said to receive his food in the stall; and this is the best way with ruminant animals. But he who finds his food for himself at the hands of Nature is better off in the open field.

Of human knowledge as a whole and in every branch of it, by far the largest part exists nowhere but on paper,—I mean, in books, that paper memory of mankind. Only a small part of it is at any given period really active in the minds of particular persons. This is due, in the main, to the brevity and uncertainty of life; but it also comes from the fact that men are lazy and bent on pleasure. Every generation attains, on its hasty passage through existence, just so much of human knowledge as it needs, and then soon disappears. Most men of learning are very superficial. Then follows a new generation, full of hope, but ignorant, and with everything to learn from the beginning. It seizes, in its turn, just so much as it can grasp or find useful on its brief journey and then too goes its way. How badly it would fare with human knowledge if it were not for the art of writing and printing! This it is that makes libraries the only sure and lasting memory of the human race, for its individual members have all of them but a very limited and imperfect one. Hence most men of learning are as loth to have their knowledge examined as merchants to lay bare their books.

Human knowledge extends on all sides farther than the eye can reach; and of that which would be generally worth knowing, no one man can possess even the thousandth part.

All branches of learning have thus been so much enlarged that he who would "do something" has to pursue no more than one subject and disregard all others. In his own subject he will then, it is true, be superior to the vulgar; but in all else he will belong to it. If we add to this that neglect of the ancient languages, which is now-a-days on the increase, and is doing away with all general education in the humanities—for a mere smattering of Latin and Greek is of no use—we shall come to have men of learning who outside their own subject display an ignorance truly bovine.

An exclusive specialist of this kind stands on a par with a workman in a factory, whose whole life is spent in making one particular kind of screw, or catch, or handle, for some particular instrument or machine, in which, indeed, he attains incredible dexterity. The specialists may also be likened to a man who lives in his own house and never leaves it. There he is perfectly familiar with everything, every little step, corner, or board; much as Quasimodo in Victor Hugo's *Notre Dame* knows the cathedral; but outside it, all is strange and unknown.

For true culture in the humanities it is absolutely necessary that a man should be many-sided and take large views; and for a man of learning in the higher sense of the word, an extensive acquaintance with history is needful. He, however, who wishes to be a complete philosopher, must gather into his head the remotest ends of human knowledge; for where else could they ever come together?

It is precisely minds of the first order that will never be specialists. For their very nature is to make the whole of existence their problem; and this is a subject upon which they will every one of them in some form provide mankind with a new revelation. For he alone can deserve the name of genius who takes the All, the Essential, the Universal, for the theme of his achievements; not he who spends his life in explaining some special relation of things one to another.

CHAPTER 48

THOMAS CARLYLE

THOMAS CARLYLE, THE SCOTTISH essayist, critic, and historian, was born at Ecclefechan in 1795. After completing his education at Edinburgh University—which he entered at fourteen—he spent several years as tutor and schoolmaster, meanwhile having entered himself as a Student of Divinity. However, after two years of preparation for the ministry of the Kirk of Scotland, to which his father had destined him, Carlyle decided that he was unsuited to the calling. During the following two years, which he spent as a private tutor, he began his literary work.

Two years after his marriage to Jane Welsh, monetary difficulties forced them to move to a dreary country estate belonging to Jane's mother. During the six years which they spent there, Carlyle produced most of his best literary work, including *Sartor Resartus*. Finally they moved to London and took a house in Chelsea where they spent the remainder of their lives. His first work written here was *The French Revolution* (a fourth of the manuscripts of which was accidentally burned and had to be rewritten). This was not a financial success, but the series of lectures which he delivered in the next few years, the books which followed (including the popular *Cromwell*), the controversial *Latter-Day Pamphlets*, and his new position as a figure in Society, started his fortunes on an upward trend. This continued during the twelve years he spent writing *Frederick the Great* and he was thus enabled to refuse Disraeli's offer of an honorary order and pension. He lingered for eight years after his important literary work was finished, passing on at the age of eighty-five.

The following quotations are taken from Carlyle's essay on *Mohammed and Mohammedanism.*

NATURE AND HEROISM

I will allow a thing to struggle for itself in this world, with any sword or tongue or implement it has, or can lay hold of. We will let it preach, and pamphleteer, and fight, and to the uttermost bestir itself, and do, beak and claws, whatsoever is in it; very sure that it will, in the long-run, conquer nothing which does not deserve to be conquered. What is better than itself, it cannot put away, but only what is worse. In this great Duel, Nature herself is umpire, and can do no wrong: the thing which is deepest-rooted in Nature, what we call *truest*, that thing and not the other will be found growing at last.

Here however,—we are to remember what an umpire Nature is; what a greatness, composure of depth and tolerance there is in her. You take wheat to cast into the Earth's bosom: your wheat may be mixed with chaff, chopped straw, barn-sweepings, dust and all imaginable rubbish; no matter: you cast it into the kind just Earth; she grows the wheat,—the whole rubbish she silently absorbs, shrouds *it* in, says nothing of the rubbish. The yellow wheat is growing there; the good Earth is silent about all the rest,—has silently turned all the rest to some benefit too, and makes no complaint about it! So everywhere in Nature! She is true and not a lie; and yet so great, and just, and motherly in her truth. She requires of a thing only that it *be* genuine of heart; she will protect it if so; will not, if not so. There is a soul of truth in all the things she ever gave harbor to. Alas, is not this the history of all highest Truth that comes or ever came into the world? The *body* of them all is imperfection, an element of light *in* darkness: to us they have to come embodied in mere Logic, in some merely *scientific* Theorem of the Universe; which *cannot* be complete; which cannot but be found, one day, *in*complete, erroneous, and so die and disappear. The body of all Truth dies; and yet in all, I say, there is a soul which never dies; which in new and ever-nobler embodiment lives immortal as man himself! It is the way with Nature. The genuine essence of Truth never dies. That it be genuine, a voice from the great Deep of Nature, there is the point at Nature's judgment seat. What *we* call pure or impure, is not with her the final question. Not how much chaff is in you; but whether you have any wheat. Pure? I might say to many a man: Yes, you are pure; pure enough; but you are chaff,—insincere hypothesis, hearsay, formality; you never were in contact with the great heart of the Universe at all; you are properly neither pure nor impure; you *are* nothing, Nature has no business with you.

It is a calumny on men to say that they are roused to heroic action by ease, hope of pleasure, recompense,—sugarplums of any kind, in this world or the next! In the meanest mortal there lies something nobler. The poor swearing soldier, hired to be shot, has his "honor of a soldier," different from drill-regulations and the shilling a day. It is not to taste sweet things, but to do noble and true things, and vindicate himself under God's Heaven as a god-made Man, that the poorest son of Adam dimly longs. Show him the way of doing that, the dullest daydrudge kindles into a hero. They wrong man greatly who say he is to be seduced by ease. Difficulty, abnegation, martyrdom, death are the *allurements* that act on the heart of man. Kindle the inner genial life of him, you have a flame that burns up all lower considerations. Not happiness, but something higher: one sees this even in the frivolous classes, with their "point of honor" and the like. Not by flattering our appetites; no, by awakening the Heroic that slumbers in every heart, can any Religion gain followers.

CHAPTER 49

AUGUSTE COMTE

ISIDORE AUGUSTE MARIE FRANCOIS Xavier Comte was born in Montpellier, France, on January 19, 1798. In 1814 he entered the École Polytechnique in Paris. Although a good student, he was expelled for mutinous conduct. He then returned to Paris and on a small annual income, retired to meditation on the social conditions of the time. Later he came under the influence of Saint-Simon, the noted philosopher, in whose association he mentally prospered. They became separated in 1824. A year later Auguste Comte was married. His marriage proved to be an unhappy one, however. In 1826 he discontinued a course of lectures which he had been delivering because of an illness of the brain. He renewed the lectures, however, in 1828. It was during the period between 1820 and 1842 that he published his famous work, *A Course of Positive Philosophy*, in which he paralleled the processes of man's thinking with the progress of the human race. There are three stages, he said, in the development of man's thought. "We begin with theological Imagination, thence we pass through metaphysical Discussion, and we end at last with positive Demonstration." This doctrine of the three stages of the development of man's thought had great influence upon subsequent philosophy.

THE POSITIVE PHILOSOPHY

A general statement of any system of philosophy may be either a sketch of a doctrine to be established, or a summary of a doctrine already established. If greater value belongs to the last, the first is still important, as characterizing from its origin the subject to be treated. In a case like the present, where the proposed study is vast and hitherto indeterminate, it is especially important that the field of research should be marked out with all possible accuracy. For this purpose, I will glance at the considerations which have originated this work, and which will be fully elaborated in the course of it.

In order to understand the true value and character of Positive Philosophy, we must take a brief general view of the progressive course of the human mind, regarded as a whole; for no conception can be understood otherwise than through its history.

From the study of the development of human intelligence, in all directions, and through all times, the discovery arises of a great fundamental law, to which it is necessarily subject, and which has a solid foundation of proof, both in the facts of our organization and in our historical experience. The law is this:—that each of our leading conceptions,—each branch of our knowledge,—passed successively through three different theoretical conditions: the Theological, or fictitious; the Metaphysical, or abstract; and the Scientific, or positive. In other words, the human mind, by its nature, employs in its progress three methods of philosophizing, the character of which is essentially different, and even radically opposed; viz., the theological method, the metaphysical and the positive. Hence arise three philosophies, or general systems of conceptions on the aggregate of phenomena, each of which excludes the others. The first is the necessary point of departure of the human understanding; and the third is its fixed and definitive state. The second is merely a state of transition.

In the theological state, the human mind, seeking the essential nature of beings, the first and final causes (the origin and purpose) of all effects,—in short, Absolute knowledge,—supposes all phenomena to be produced by the immediate action of supernatural beings.

In the metaphysical state, which is only a modification of the first, the mind supposes, instead of supernatural beings, abstract forces, veritable entities (that is, personified abstractions) inherent in all beings, and capable of producing all phenomena. What is called the explanation of phenomena is, in this stage, a mere reference of each to its proper entity.

In the final, the positive state, the mind has given over the vain search after Absolute notions, the origin and destination of the universe, and the causes of phenomena, and applies itself to the study of their laws,—that is, their invariable relations of succession and resemblance. Reasoning and observation, duly combined, are the means of this knowledge. What is now understood when we speak of an explanation of facts is simply the establishment of a connection between single phenomena and some general facts, the number of which continually diminishes with the progress of science.

The Theological system arrived at the highest perfection of which it is capable when it substituted the providential action of a single Being for the varied operations of the numerous divinities which had been before imagined. In the same way, in the last stage of the Metaphysical system, men substitute one great entity (Nature) as the cause of all phenomena, instead of the multitude of entities at first supposed. In the same way, again, the ultimate perfection of the Positive system would be (if such perfection could be hoped for) to represent all phenomena as particular aspects of a single general fact;—such as Gravitation, for instance.

The importance of the working of this general law will be established hereafter. At present, it must suffice to point out some of the grounds of it.

There is no science which, having attained the positive stage, does not bear marks of having passed through the others. Some time since it was (whatever it might be) composed, as we can now perceive, of metaphysical abstractions; and, further back in the course of time, it took its form from theological conceptions. We shall have only too much occasion to see, as we proceed, that our most advanced sciences still bear very evident marks of the two earlier periods through which they have passed.

The progress of the individual mind is not only an illustration, but an indirect evidence of that of the general mind. The point of departure of the individual and of the race being the same, *the phases of the mind of a man correspond to the epochs of the mind of the race.* Now, each of us is aware, if he looks back upon his own history, that he was a theologian in his childhood, a metaphysician in his youth, and a natural philosopher in his manhood. All men who are up to their age can verify this for themselves.

CHAPTER 50

HONORÉ DE BALZAC

HONORÉ DE BALZAC, THE famous French novelist and mystic, was born in Tours, France, May 20, 1799. He was schooled in the common French schools until nine and then was placed by his parents in a strict and gloomy ecclesiastical boarding school. His parents required him to work as a legal clerk while he was studying law at the University, but he rebelled upon graduation and retired to a garret where on a meager income he nearly starved. It was while in this garret that some of his greatest literary contributions were made. He worked, living on small rations, as long as sixteen hour stretches, daily, with merely a little relaxation between. He seemed to have read but little in the latter part of his life and, therefore, must have read much in his youth. His lack of appreciation for money caused him much embarrassment, and he vacillated from riches to poverty.

As a biographer says, in part of his literary works "He seemed accustomed to create in a fashion which is not so much of the actual world as of some other, possible but not actual—"

He finally became very weak, constitutionally, due to his impoverished condition and overwork, and experienced transition August 17, 1850. Some who were pallbearers are now considered renowned geniuses, such as Victor Hugo and Alexander Dumas.

THE INFINITE

If material science be the end and object of human effort, tell me, both of you, would societies—those great centers where men congregate—would they perpetually be dispersed? If civilization were the object of our Species, would intelligence perish? Would it continue purely individual? The grandeur of all nations that were truly great was based on exceptions; when the exception ceased their power died. If such were the End-all, Prophets, Seers, and Messengers of God would have lent their hand to Science rather than have given it to Belief. Surely they would have quickened your brains sooner than have touched your hearts! But no; one and all they came to lead the nations back to God; they proclaimed the sacred Path in simple words that showed the way to heaven; all were wrapped in love and faith, all were inspired by that WORD which hovers above the inhabitants of earth, enfolding them, inspiriting them, uplifting them; none were prompted by any human interest. Your great geniuses, your poets, your kings, your learned men are engulfed with their cities; while the names of these good pastors of humanity, ever blessed, have survived all cataclysms. . . .

When you call God a Creator, you dwarf Him. He did not create, as you think He did, plants or animals or stars. Could He proceed by a variety of means? Must He not act by unity of composition? Moreover, He gave forth principles to be developed, according to His universal law, at the will of the surroundings in which they were placed. Hence a single substance and motion, a single plant, a single animal, but correlations everywhere. In fact, all affinities are linked together by contiguous similitudes; the life of the worlds is drawn toward the centers by famished aspirations, as you are drawn by hunger to seek food.

To give you an example of affinities linked to similitudes (a secondary law on which the creations of your thought are based), music, that celestial art, is the working out of this principle; for is it not a complement of sounds harmonized by number? Is not sound a modification of air, compressed, diluted, echoed? You know the composition of air,—oxygen, nitrogen, and carbon. As you cannot obtain sound from the void, it is plain that music and the human voice are the result of organized chemical substances, which put themselves in unison with the same substances prepared within you by your thought, coordinated by means of light, the great nourisher of your globe. Have you ever meditated on the masses of nitre deposited by the snow, have you ever observed a thunder-storm and seen the plants breathing in from the air about them the metal it contains, without concluding that the sun has fused and distributed the subtle essence which nourishes all things here below? . . .

Can God abase himself to you? Is it not for you to rise to Him? If human reason finds the ladder of its own strength too weak to bring God down to it, is it not evident that you must find some other path to reach Him? That Path is in ourselves. The Seer and the Believer find eyes within their souls more piercing far than eyes that probe the things on earth,—they see the Dawn. Hear this truth: Your science, let it be ever so exact, your meditations, however bold, your noblest lights, are Clouds. Above, above is the Sanctuary whence the true Light flows. . . .

We have also admitted that Matter and Spirit are two creations which do not comprehend each other; that the spiritual world is formed of infinite relations to which the finite material world has given rise; that if no one on earth is able to identify himself by the power of his spirit with the great-whole of terrestrial creations, still less is he able to rise to the knowledge of the relations which the spirit perceives between these creations.

We might end the argument here in one word, by denying you the faculty of comprehending God, just as you deny to the pebbles of the fiord the faculties of counting and of seeing each other. How do you know that the stones themselves do not deny the existence of man, though man makes use of them to build his houses? There is one fact that appals you,—the Infinite; if you feel it within you, why will you not admit the consequences? Can the finite have a perfect knowledge of the infinite? If you cannot perceive those relations which, according to your own admission, are infinite, how can you grasp a sense of the far-off end to which they are converging? Order, the revelation of which is one of your needs, being infinite, can your limited reason apprehend it? Do not ask why man does not comprehend that which he is able to perceive, for he is equally able to perceive that which he does not comprehend. If I prove to you that your mind ignores that which lies within its compass, will you grant that it is impossible for it to conceive whatever is beyond it? This being so, am I not justified in saying to you: "One of the two propositions under which God is annihilated before the tribunal of our reason must be true, the other is false. Inasmuch as creation exists, you feel the necessity of an end, and that end should be good, should it not? Now, if Matter terminates in man by intelligence, why are you not satisfied to believe that the end of human intelligence is the Light of the higher spheres, where alone an intuition of that God who seems to you so insoluble a problem is obtained? The species which are beneath you have no conception of the universe, and you have; why should there not be other species above you more intelligent than your own? Man ought to be better informed than he is about himself before he spends his strength in measuring God. Before attacking the stars that light us, and the higher certainties, ought he not to understand the certainties which are actually about him?

CHAPTER 51

RALPH WALDO EMERSON

RALPH WALDO EMERSON Is a true American literary product. He was born May 25, 1803, and was a son of a prominent Unitarian minister in Boston, Massachusetts. He was educated at the Boston Latin School and later at Harvard University. He then taught school for some time, and in 1825 returned to Harvard to study divinity. He was married in 1829, and that same year became minister of a Boston church. In 1831 his wife died. The next year, scruples about administering the Lord's Supper caused him to give up his church. In sadness and poor health, he traveled extensively through Europe, laying the foundations of a life-time friendship with Coleridge, Landor, Wordsworth, and Carlyle, whom he visited. Returning to America, he took up lecturing, and continued for nearly forty years to use this method of expressing his ideas on religion, politics, literature, and philosophy.

He aided in founding the Transcendentalist periodical, *The Dial.* He arranged for the American publication of Carlyle's great *Sartor Resartus* at a time when it was being refused by British publishers. He wrote numerous volumes of essays, addresses, and poems.

Later, he again went to Europe and lectured extensively in England and Scotland. He died in Concord on April 7, 1882. At the time of his death he was recognized as an outstanding writer and thinker of America. His frank departure from orthodoxy had a tendency to make him unpopular in certain circles. He is referred to, even today, as the greatest of American thinkers.

SELF-RELIANCE

I read the other day some verses written by an eminent painter which were original and not conventional. Always the soul hears an admonition in such lines, let the subject be what it may. The sentiment they instill is of more value than any thought they may contain. To believe your own thought, to believe that what is true for you in your private heart is true for all men,— that is genius. Speak your latent conviction, and it shall be the universal sense; for always the inmost becomes the outmost—and our first thought is rendered back to us by the trumpets of the Last Judgment. Familiar as the voice of the mind is to each, the highest merit we ascribe to Moses, Plato and Milton is that they set at naught books and traditions, and spoke not what men, but what they thought. A man should learn to detect and watch that gleam of light which flashes across his mind from within more than the lustre of the firmament of bards and sages. Yet he dismisses without notice this thought, because it is his. In every work of genius we recognize our own rejected thoughts; they come back to us with a certain alienated majesty. Great works of art have no more affecting lesson for us than this. They teach us to abide by our spontaneous impression with a good-humored inflexibility greater than most when the whole cry of voices is on the other side. Else to-morrow a stranger will say with masterly good sense precisely what we have thought and felt all the time, and we shall be forced to take with shame our own opinion from another.

There is a time in every man's education when he arrives at the conviction that envy is ignorance; that imitation is suicide; that he must take himself for better or worse as his portion; that though the wide universe is full of good, no kernel of nourishing corn can come to him but through his toil bestowed on that plot of ground which is given to him to till. The power which resides in him is new in nature, and none but he knows what that is which he can do, nor does he know until he has tried. Not for nothing one face, one character, one fact, makes much impression on him, and another none. It is not without preestablished harmony, this sculpture in the memory. The eye was placed where one ray should fall, that it might testify of the particular ray. Bravely let him speak the utmost syllable of his confession. We but half express ourselves, and are ashamed of that divine idea which each of us represents. It may be safely trusted as proportionate and of good issues, so it be faithfully imparted, but God will not have his work made manifest by cowards. It needs a divine man to exhibit anything divine. A man is relieved and gay when he has put his heart into his work and done his best; but what he has said or done otherwise shall give him no peace. It is a deliverance which does not deliver. In the attempt his genius deserts him; no muse befriends; no invention, no hope.

Trust thyself: every heart vibrates to that iron string. Accept the place the divine providence has found for you, the society of your contemporaries, the connexion of events. Great men have always done so, and confided themselves childlike to the genius of their age, betraying their perception that the Eternal was stirring at their heart, working through their hands, predominating in all their being. And we are now men, and must accept in the highest mind the same transcendent destiny; and not pinched in a corner, not cowards fleeing before a revolution, but redeemers and benefactors, pious aspirants to be noble clay under the Almighty effort let us advance on Chaos and the Dark.

What pretty oracles nature yields us on this text in the face and behavior of children, babes, and even brutes. That divided and rebel mind, that distrust of a sentiment because our arithmetic has computed the strength and means opposed to our purpose, these have not. Their mind being whole, their eye is as yet unconquered, and when we look in their faces, we are disconcerted. Infancy conforms to nobody; all conform to it; so that one babe commonly makes four or five out of the adults who prattle and play to it. So God has armed youth and puberty and manhood no less with its own piquancy and charm, and made it enviable and gracious and its claims not to be put by, if it will stand by itself. Do you think the youth has no force, because he cannot speak to you and me? Hark! in the next room who spoke so clear and emphatic? It seems he knows how to speak to his contemporaries. Good Heaven! it is he! it is that very lump of bashfulness and phlegm which for weeks has done nothing but eat when you were by, and rolls out these words like bell-strokes. It seems he knows how to speak to his contemporaries. Bashful or bold then, he will know how to make us seniors very unnecessary.

CHAPTER 52

LUDWIG ANDREAS FEUERBACH

LUDWIG ANDREAS FEUERBACH, A German philosopher, was born at Landshut, Bavaria, July 28, 1804. After devoting two years to the study of theology at the University of Heidelberg, he was attracted to Berlin in 1824 for the purpose of hearing Hegel. Soon afterward he abandoned theology to devote himself entirely to philosophy, of which in 1828 he became an instructor in the University of Erlangen. In a few years, however, he quitted his academical activities on account of offense given by his first book, published anonymously in 1830, in which he attacks the belief in personal immortality and advocates the Spinozistic immortality of reabsorption in nature. During the next few years he published three works on portions of the history of philosophy, treating the period from Bacon to Spinoza, and the theories of Leibnitz and Bayle. Such historical research only paved the way to a critical investigation into the nature of religion and its relation to philosophy, which resulted in his most celebrated work, *The Essence of Christianity*, in 1841. Feuerbach herein denies to God any existence except as an idealized object of human consciousness, thus making the conception of God merely the projection by man of his own ideal into the objective world. All authority above man is regarded as a delusion proceeding from man himself, and the highest good is explained as consisting in resemblance to that ideal humanity which man creates for himself and worships as God.

Feuerbach's greatest influence has been upon the anti-Christian theologians who like himself had passed from Hegelianism to a form of naturalism. However, his work was too deliberately unsystematic ever to make him a power in philosophy. In spite of his philosophical materialism, he was a man of high ideals, and at the time of his death in 1872 he had made numerous friends both at home and abroad. Below are some interesting excerpts from his most important work entitled *The Essence of Christianity*, and although his philosophy may not be acceptable to the reader, a careful analysis of his viewpoints and contentions will be both beneficial and profitable.

REASON, WILL, AFFECTION

What, then, is the nature of man, of which he is conscious, or what constitutes the specific distinction, the proper humanity of man? Reason, Will, Affection. To a complete man belong the power of Thought, the power of Will, the power of Affection. The power of Thought is the light of the intellect, the power of Will is energy of character, the power of Affection is love. Reason, love, force of will are perfections—the perfections of the human being—nay, more, they are absolute perfections of being. To will, to love, to think, are the highest powers, are the absolute nature of man as man, and the basis of his existence. Man exists to think, to love, to will. Now that which is the end, the ultimate aim, is also the true basis and principle of a being. But what is the end of reason? Reason. Of love? Love. Of Will? Freedom of the Will. We think for the sake of thinking, loving, willing existence. That alone is true, perfect, divine, which exists for its own sake. But such is love, such is reason, such is will. The divine trinity in man, above the individual man, is the unity of reason, love, will. Reason, Will, Love, are not powers which a man possesses, for he is nothing about them; he is what he is only by them; they are the constituent elements of his nature, which he neither has nor makes, the animating, determining, governing powers—divine, absolute powers—to which he can oppose no resistance.

How can the feeling man resist feeling, the loving one love, the rational one reason? Who has not experienced the overwhelming power of melody? And what else is the power of melody but the power of feeling? Music is the language of feeling; melody is audible feeling—feeling communicating itself. Who has not experienced the power of love, or at least heard of it? Which is the stronger—love or the individual man? It is man possesses love, or is it not much rather love that possesses man? When love impels a man to suffer death even joyfully for the beloved one, is this death-conquering power his own individual power, or is it not rather the power of love? And who that ever truly thought has not experienced that quiet, subtle power—the power of thought? When you sinkest into deep reflection, forgetting thyself and what is around thee, dost thou govern reason, or is it not reason which governs and absorbs thee? Scientific enthusiasm—is it not the most glorious triumph of intellect over thee? The desire of knowledge—is it not a simply irresistible and an all-conquering power? And when thou suppressest a passion, renouncest a habit, achievest a victory over thyself, is this victorious power thine own personal power, or is it not rather the energy of will, the force of morality, which seizes the mastery of thee, and fills thee with indignation against thyself and thine individual weakness?

Man's Nature His Sole Object of Consciousness

Man is nothing without an object. The great models of humanity, such men as reveal to us what man is capable of, have attested the truth of this proposition by their lives. They had only one dominant passion—the realization of the aim which was the essential object of their activity. But the object to which a subject essentially, necessarily relates is nothing else than this subject's own, but objective, nature. If it be an object common to several individuals of the same species, but under various conditions, it is still, at least as to the form under which it presents itself to each of them according to their respective modifications, their own, but objective, nature. . . .

In the object which he contemplates, therefore, man becomes acquainted with himself; consciousness of the objective is the self-consciousness of man. We know the man by the object, by his conception of what is external to himself; in it his nature becomes evident; this object is his manifested nature, his true objective ego. And this is true, not merely of spiritual, but also of sensuous objects. Even the objects which are most remote from man, because they are objects to him, and to the extent that they are so, are revelations of human nature. That he sees them and so sees them is an evidence of his own nature. The animal is sensible only of the beam which immediately affects life; while man perceives the ray, to him physically indifferent, of the remoter star. Man alone has purely intellectual, disinterested joys and passions; the eye of man alone keeps theoretic festivals. . . .

The absolute to man is his own nature. The power of the object over him is therefore the power of his own nature. Thus the power of the object of feeling is the power of feeling itself; the power of the object of the intellect is the power of the intellect itself; the power of the object of the will is the power of the will itself. The man who is affected by musical sounds is governed by feeling; by the feeling, that is, which finds its corresponding element in musical sounds. But it is not melody as such, it is only melody pregnant with meaning and emotion, which has power over feeling. Feeling is only acted on by that which conveys feeling, i.e., by itself, its own nature. Thus also the will; thus, and infinitely more, the intellect. Whatever kind of object, therefore, we are at any time conscious of, we are always at the same time conscious of our own nature; we can affirm nothing without affirming ourselves. And since to will, to feel, to think, are perfections, essences, realities, it is impossible that intellect, feeling, and will should feel or perceive themselves as limited, finite and nothingness are identical; finiteness is only a power, i.e., as worthless, as nothing. For finiteness is a euphemism for nothingness. Finiteness is the metaphysical, the theoretical—nothingness the pathological, practical expression. What is finite to the understanding is nothing to the heart.

But it is impossible that we should be conscious of will, feeling, and intellect as finite powers, because every perfect existence, every original power and essence, is the immediate verification and affirmation of itself. It is impossible to love, will, or think, without perceiving these activities to be perfections—impossible to feel that one is a loving, willing, thinking being without experiencing an infinite joy therein. Consciousness consists in a being becoming objective to itself; hence it is nothing apart, nothing distinct from the being which is conscious of itself. How could it otherwise become conscious of itself; it is, therefore, impossible to become conscious of a perfection as an imperfection, impossible, to feel feeling limited, to think thought limited.

CHAPTER 53

CHARLES DARWIN

CHARLES ROBERT DARWIN WAS born in Shrewsbury, England, on February 12, 1809. He was the grandson of the illustrious Erasmus Darwin. He studied at Edinburgh and Cambridge and he graduated in 1831. He was engaged as a naturalist for the scientific expedition of the Beagle, which went around the world from 1831 to 1836. At first he was but a collector, but this expedition caused him to become an investigator, and through it he gained his first ideas of evolution. In the year 1838 he read Malthus on the increase of population. This work influenced him greatly. His theory of natural selection first appears in his notes about 1844. It was about this time that he began his work entitled *On the Origin of Species by Means of Natural Selection* which he finished in 1859.

For quite some years past, the theories of evolution and natural selection have been looked upon as hostile to religion, as contrary to scripture. This view is mainly taken by those who either have no appreciation of the real significance of the theories, or who interpret the Bible literally. The doctrine and theory of evolution is in accord with natural law and with Divine principle. The old orthodox conception that man is the result of special creation, and sprang into existence as a complete species, cannot be supported by the facts of nature. The true facts that man has evolved to his present status through his own efforts and in accordance with natural law adds to the glory of the Creator, rather than detracting from it. The basic idea of Darwin is that organisms produce more rapidly than there is sustenance, and the different organisms vary, and because of this variation only those that are stronger and better qualified to survive do so.

NATURAL SELECTION

We shall best understand the probable course of natural selection by taking the case of a country undergoing some slight physical change, for instance, of climate. The proportional numbers of its inhabitants will almost immediately undergo a change, and some species will probably become extinct. We may conclude, from what we have seen of the intimate and complex manner in which the inhabitants of each country are bound together, that any change in the numerical proportions of the inhabitants, independently of the change of climate itself, would seriously affect the others. If the country were open on its borders, new forms would certainly immigrate, and this would likewise seriously disturb the relations of some of the former inhabitants. Let it be remembered how powerful the influence of a single introduced tree or mammal has been shown to be. But in the case of an island, or of a country partly surrounded by barriers, into which new and better adapted forms could not freely enter, we should then have places in the economy of nature which would assuredly be better filled up, if some of the original inhabitants were in some manner modified; for, had the area been open to immigration, these same places would have been seized on by intruders. In such cases, slight modifications, which in any way favoured the individuals of any species, by better adapting them to their altered conditions, would tend to be preserved; and natural selection would have free scope for the work of improvement.

We have good reason to believe, as shown in the first chapter, that changes in the conditions of life give a tendency to increased variability; and in the foregoing cases the conditions have changed, and this would manifestly be favourable to natural selection, by affording a better chance of the occurrence of profitable variations. Unless such occur, natural selection can do nothing. Under the term of "variations," it must never be forgotten that mere individual differences are included. As man can produce a great result with his domestic animals and plants by adding up in any given direction individual differences, so could natural selection, but far more easily from having incomparably longer time for action. Nor do I believe that any great physical change, as of climate, or any unusual degree of isolation to check immigration, is necessary in order that new and unoccupied places should be left for natural selection to fill up by improving some of the varying inhabitants. For as all the inhabitants of each country are struggling together with nicely balanced forces, extremely slight modifications in the structure or habits of one species would often give it an advantage over others; and still further modifications of the same kind would often still further increase the advantage, as long as the species continued under the same conditions of life and profited by similar means of subsistence and defense. No country can be named in which all the native inhabitants are now so perfectly adapted to each other and to the physical conditions under which they live, that none of them could be still better adapted or improved; for in all countries, the natives have been so far conquered by naturalized productions, that they have allowed some foreigners to take firm possession of the land. And as foreigners have thus in every country beaten some of the natives, we may safely conclude that the natives might have been modified with advantage, so as to have better resisted the intruders.

As man can produce, and certainly has produced, a great result by his methodical and unconscious means of selection, what may not natural selection effect? Man can act only on external and visible characters: Nature, if I may be allowed to personify the natural preservation or survival of the fittest, cares nothing for appearances, except in so far as they are useful to any being. She can act on every internal organ, on every shade of constitutional difference, on the whole machinery of life. Man selects only for his own good: Nature only for that of the being which she tends. Every selected character is fully exercised by her, as is implied by the fact of their selection. Man keeps the natives of many climates in the same country; he seldom exercises each selected character in some peculiar and fitting manner; he feeds a long and a short-beaked pigeon on the same food; he does not exercise a long-backed or long-legged quadruped in any peculiar manner; he exposes sheep with long and short wool to the same climate. He does not allow the most vigorous males to struggle for the females. He does not rigidly destroy all inferior animals, but protects during each varying season, as far as lies in his power, all his production. He often begins his selection by some half-monstrous form; or at least by some modification prominent enough to catch the eye or to be plainly useful to him. Under nature, the slightest differences of structure or constitution may well turn the nicely-balanced scale in the struggle for life, and so be preserved. How fleeting are the wishes and efforts of man! how short his time! and consequently how poor will be his results, compared with those accumulated by Nature during whole geological periods! Can we wonder, then, that Nature's productions should be far "truer" in character than man's productions; that they should be infinitely better adapted to the most complex conditions of life, and should plainly bear the stamp of far higher workmanship?

It may metaphorically be said that natural selection is daily and hourly scrutinizing, throughout the world, the slightest variations; rejecting those that are bad, preserving and adding up all that are good; silently and sensibly working, whenever and wherever opportunity offers, at the improvement of each organic being in relation to its organic and inorganic conditions of life. We see nothing of these slow changes in progress, until the hand of time has marked the lapse of ages, and then so imperfect is our view into long-past geological ages, that we see only that the forms of life are now different from what they formerly were.

In order that any great amount of modification should be effected in a species, a variety when once formed must again, perhaps after a long interval of time, vary or present individual differences of the same favourable nature as before; and these must be again preserved, and so onwards step by step. Seeing that individual differences of the same kind perpetually recur, this can hardly be considered as an unwarrantable assumption. But whether it is true, we can judge only by seeing how far the hypothesis accords with and explains the general phenomena of nature. On the other hand, the ordinary belief that the amount of possible variation is a strictly limited quantity is likewise a simple assumption.

Although natural selection can act only through and for the good of each being, yet characters and structures, which we are apt to consider as of very trifling importance, may thus be acted on. When we see leaf-eating insects green, and bark-feeders mottled gray; the Alpine ptarmigan white in winter, the red-grouse the colour of heather, we must believe that these tints are of service to these birds and insects in preserving them from danger. Grouse, if not destroyed at some period of their lives, would increase in countless numbers; they are known to suffer largely from birds of prey; and hawks are guided by eyesight to their prey—so much so, that on parts of the Continent persons are warned not to keep white pigeons, as being the most liable to destruction. Hence natural selection might be effective in giving the proper colour to each kind of grouse, and in keeping that colour, when once acquired, true and constant. Nor ought we to think that the occasional destruction of an animal of any particular colour would produce little effect: we should remember how essential it is in a flock of white sheep to destroy a lamb with the faintest trace of black. We have seen how the colour of the hogs, which feed on the "paint-root" in Virginia, determines whether they shall live or die. In plants, the down on the fruit and the colour of the flesh are considered by botanists as characters of the most trifling importance: yet we hear from an excellent horticulturist, Downing, that in the United States smooth-skinned fruits suffer far more from a beetle, a Curculio, than those with down; that purple plums suffer far more from a certain disease than yellow plums; whereas another disease attacks yellow-fleshed peaches far more than those with other coloured flesh. If, with all the aids of arts, these slight differences make a great difference in cultivating the several varieties, assuredly, in a state of nature, where the trees would have to struggle with other trees and with a host of enemies, such differences would effectually settle which variety, whether a smooth or downy, a yellow or purple-fleshed fruit, should succeed.

In looking at many small points of difference between species, which, as far as our ignorance permits us to judge, seem quite unimportant, we must not forget that climate, food, etc., have no doubt produced some direct effect. It is also necessary to bear in mind that, owing to the law of correlation, when one part varies, and the variations are accumulated through natural selection, other modifications, often of the most unexpected nature, will ensue.

As we see that those variations which, under domestication, appear at any particular period of life, tend to reappear in the offspring at the same period; for instance, in the shape, size, and flavour of the seeds of the many varieties of our culinary and agricultural plants; in the caterpillar and cocoon stages of the varieties of the silkworm; in the eggs of poultry, and in the colour of the down of their chickens; in the horns of our sheep and cattle when nearly adult; so in a state of nature natural selection will be enabled to act on and modify organic beings at any age, by the accumulation of variations profitable at that age, and by their inheritance at a corresponding age. If it profit a plant to have its seeds more and more widely disseminated by the wind, I can see no greater difficulty in this being effected through natural selection, than in the cotton planter increasing and improving by selection the down in the pods on his cotton trees. Natural selection may modify and adapt the larva of an insect to a score of contingencies, wholly different from those which concern the mature insect; and these modifications may effect, through correlation, the structure of the adult. So, conversely, modifications in the adult may affect the structure of the larva; but in all cases natural selection will insure that they shall not be injurious: for if they were so, the species would become extinct.

Natural selection will modify the structure of the young in relation to the parent, and of the parent in relation to the young. In social animals it will adapt the structure of each individual for the benefit of the whole community; if the community profits by the selected change. What natural selection cannot do, is to modify the structure of one species, without giving it any advantage, for the good of another species; and though statements to this effect may be found in works of natural history, I cannot find one case which will bear investigation. A structure used only once in an animal's life, if of high importance to it, might be modified to any extent by natural selection; for instance, the great jaws possessed by certain insects, used exclusively for opening the cocoon—or the hard tip to the beak of unhatched birds, used for breaking the egg. It has been asserted, that of the best short-beaked tumbler-pigeons a greater number perish in the egg than are able to get out of it; so that fanciers assist in the act of hatching. Now if nature had to make the beak of a full-grown pigeon very short for the bird's own advantage, the process of modification would be very slow, and there would be simultaneously the most rigorous selection of all the young birds within the egg, which had the most powerful and hardest beaks, for all with weak beaks would inevitably perish; or, more delicate and more easily broken shells might be selected, the thickness of the shell being known to vary like every other structure.

It may be well here to remark that with all beings there must be much fortuitous destruction, which can have little or no influence on the course of natural selection. For instance a vast number of eggs or seeds are annually devoured, and these could be modified through natural selection only if they varied in some manner which protected them from their enemies. Yet many of these eggs or seeds would perhaps, if not destroyed, have yielded individuals better adapted to their conditions of life than any of those which happened to survive. So again a vast number of mature animals and plants, whether or not they be the best adapted to their conditions, must be annually destroyed by accidental causes, which would not be in the least degree mitigated by certain changes of structure or constitution which would in other ways be beneficial to the species. But let the destruction of the adults be ever so heavy, if the number which can exist in any district be not wholly kept down by such causes,—or again let the destruction of eggs or seeds be so great that only a hundredth or a thousandth part are developed,—yet of those which do survive, the best adapted individuals, supposing that there is any variability in a favourable direction, will tend to propagate their kind in larger numbers than the less well adapted. If the numbers be wholly kept down by the causes just indicated, as will often have been the case, natural selection will be powerless in certain beneficial directions; but this is no valid objection to its efficiency at other times and in other ways; for we are far from having any reason to suppose that many species ever undergo modification and improvement at the same time in the same area.

Sexual Selection

Inasmuch as peculiarities often appear under domestication in one sex and become hereditarily attached to that sex, so no doubt it will be under nature. Thus it is rendered possible for the two sexes to be modified through natural selection in relation to different habits of life, as is sometimes the case; or for one sex to be modified in relation to the other sex, as commonly occurs. This leads me to say a few words on what I have called Sexual Selection. This form of selection depends, not on a struggle for existence in relation to other organic beings or to external conditions, but on a struggle between the individuals of one sex, generally the males, for the possession of the other sex. The result is not death to the unsuccessful competitor, but few or no offspring. Sexual selection is, therefore, less rigorous than natural selection. Generally, the most vigorous males, those which are best fitted for their places in nature, will leave most progeny. But in many cases, victory depends not so much on general vigour, as on having special weapons, confined to the male sex. A hornless stag or spurless cock would have a poor chance of leaving numerous offspring. Sexual selection, by always allowing the victor to breed, might surely give indomitable courage, length to the spur, and strength to the wing to strike in the spurred leg, in nearly the same manner as does the brutal cockfighter by the careful selection of his best cocks. How low in the scale of nature the law of battle descends, I know not; male alligators have been described as fighting, bellowing, and whirling round, like Indians in a war-dance, for the possession of the females; male salmons have been observed fighting all day long; male stag-beetles sometimes bear wounds from the huge mandibles of other males; the males of certain hymenopterous insects have been frequently seen by that inimitable observer, M. Fabre, fighting for a particular female who sits by, an apparently unconcerned beholder of the struggle, and then retires with the conqueror. The war is, perhaps, severest between the males of polygamous animals, and these seem oftenest provided with special weapons. The males of carnivorous animals are already well armed; though to them and to others, special means of defence may be given through means of sexual selection, as the mane of the lion, and the hooked jaw to the male salmon; for the shield may be as important for victory as the sword or spear.

Amongst birds, the contest is often of a more peaceful character. All those who have attended to the subject believe that there is the severest rivalry between the males of many species to attract, by singing, the females. The rock-thrush of Guiana, birds of paradise, and some others, congregate; and successive males display with the most elaborate care, and show off in the best manner, their gorgeous plumage; they likewise perform strange antics before the females, which, standing by as spectators, at last choose the most attractive partner. Those who have closely attended to birds in confinement well know that they often take individual preferences and dislikes: thus Sir R. Heron has described how a pied peacock was eminently attractive to all his hen birds. I cannot here enter on the necessary details; but if man can in a short time give beauty and an elegant carriage to his bantams, according to his standard of beauty, I can see no good reason to doubt that female birds, by selecting, during thousands of generations, the most melodious or beautiful males, according to their standard of beauty, might produce a marked effect. Some well-known laws, with respect to the plumage of male and female birds, in comparison with the plumage of the young, can partly be explained through the action of sexual selection on variations occurring at different ages, and transmitted to the males alone or to both sexes at corresponding ages; but I have not space here to enter on this subject.

Thus it is, as I believe, that when the males and females of any animal have the same general habits of life, but differ in structure, colour, or ornament, such differences have been mainly caused by sexual selection: that is, by individual males having had, in successive generations, some slight advantage over other males, in their weapons, means of defence, or charms, which they have transmitted to their male offspring alone. Yet, I would not wish to attribute all sexual differences to this agency: for we see in our domestic animals peculiarities arising and becoming attached to the male sex, which apparently have not been augmented through selection by man. The tuft of hair on the breast of the wild turkey-cock cannot be of any use, and it is doubtful whether it can be ornamental in the eyes of the female bird;—indeed, had the tuft appeared under domestication, it would have been called a monstrosity.

CHAPTER 54

SIR WILLIAM CROOKES

SIR WILLIAM CROOKES WAS born in London in 1832 and died in 1919. In 1851 he gave himself to original research in chemistry. Later, in 1859, he founded a publication known as *The Chemical News,* and in 1884 he also became editor of *The Quarterly Journal of Science.* He was both practical and theoretical. He went beyond the limits prescribed by his particular field of science and explored the realm of metaphysics. His practical research made him an authority on sewage, beet sugar, dyeing, and calico printing. He was the inventor of the famous Crookes' tube, the parent of the radio vacuum tube. His early research in this field was one of the contributing factors toward electrical therapeutics and radio telephony. From his experiments, Roentgen established the foundation for the invention of the X-ray. He devoted himself extensively to the theory of the composition of atoms. His great interest in metaphysics and particularly in psychic phenomena may be understood from these excerpts from his writings on telepathy.

TELEPATHY

The task I am called upon to perform today is to my thinking by no means a merely formal or easy matter. It fills me with deep concern to give an address, with such authority as a president's chair confers, upon a science which though still in a purely nascent stage, seems to me at least as important as any other science whatever. Psychical science, as we here try to pursue it, is the embryo of something which in time may dominate the whole world of thought. This possibility—nay, probability—does not make it the easier to me now. Embryonic development is apt to be rapid and interesting; yet the prudent man shrinks from dogmatising on the egg until he has seen the chicken.

Nevertheless, I desire, if I can, to say a helpful word. And I ask myself what kind of helpful word. Is there any connection between my old-standing interest in psychical problems and such original work as I may have been able to do in other branches of science?

I think there is such a connection—that the most helpful quality which has aided me in psychical problems and has made me lucky in physical discoveries (sometimes of rather unexpected kinds) has simply been my knowledge—my vital knowledge, if I may so term it— of my own ignorance.

Most students of nature sooner or later pass through a process of writing off a large percentage of their supposed capital of knowledge as a merely illusory asset. As we trace more accurately certain familiar sequences of phenomena we begin to realize how closely these sequences, or laws, as we call them, are hemmed round by still other laws of which we can form no notion. With myself this writing off of illusory assets has gone rather far and the cobweb of supposed knowledge has been pinched (as some one has phrased) into a particularly small pill.

Telepathy, the transmission of thought and images directly from one mind to another without the agency of the recognized organs of sense, is a conception new and strange to science. To judge from the comparative slowness with which the accumulated evidence of our society penetrates the scientific world, it is, I think, a conception even scientifically repulsive to many minds. We have supplied striking experimental evidence; but few have been found to repeat our experiment. We have offered good evidence in the observation of spontaneous cases, as apparitions at the moment of death and the like, but this evidence has failed to impress the scientific world in the same way as evidence less careful and less coherent has often done before. Our evidence is not confronted and refuted; it is shirked and evaded as though there were some great *a priori* improbability which absolved the world of science from considering it. I at least see no *a priori* improbability whatever. Our alleged facts might be true in all kinds of ways without contradicting any truth already known. I will dwell now on only one possible line of explanation, not that I see any way of elucidating all the new phenomena I regard as genuine, but because it seems probable I may shed a light on some of those phenomena.

All the phenomena of the universe are presumably in some way continuous; and certain facts, plucked as it were from the very heart of nature, are likely to be of use in our gradual discovery of facts which lie deeper still.

Let us, then, consider the vibrations we trace, not only in solid bodies, but in the air, and in a still more remarkable manner in the ether.

These vibrations differ in their velocity and in their frequency. That they exist, extending from one vibration to two thousand millions of millions vibrations per second, we have good evidence.

That they subserve the purpose of conveying impressions from outside sources of whatever kind to living organisms may be fully recognized. . . .

Ordinarily we communicate intelligence to each other by speech. I first call up in my own brain a picture of a scene I wish to describe, and then, by means of an orderly transmission of wave vibrations set in motion by my vocal chords through the material atmosphere, a corresponding picture is implanted in the brain of anyone whose ear is capable of receiving such vibrations. If the scene I wish to impress on the brain of the recipient is of a complicated character, or if the picture of it in my own brain is not definite, the transmission will be more or less imperfect; but if I wish to get my audience to picture to themselves some very simple object, such as a triangle or a circle, the transmission of ideas will be well-nigh perfect, and equally clear to the brains of both transmitter and recipient. Here we use the vibrations of the material molecules of the atmosphere to transmit intelligence from one brain to another.

In the newly discovered Roentgen rays we are introduced to an order of vibrations of extremest minuteness as compared with the most minute waves with which we have hitherto been acquainted, and of dimensions comparable with the distance between the centers of the atoms of which the material universe is built up; and there is no reason to suppose that we have here reached the limit of frequency. Waves of this character cease to have many of the properties associated with light waves. They are produced in the same ethereal medium, and are probably propagated with the same velocity as light, but here the similarity ends. They cannot be regularly reflected from polished surfaces; they have not been polarized; they are not refracted on passing from one medium to another of different density, and they penetrate considerable thickness of substances opaque to light with the same ease with which light passes through glass. It is also demonstrated that these rays, as generated in the vacuum tube, are not homogeneous, but consist of bundles of different wave-lengths, analogous to what would be differences of color could we see them as light. Some pass easily through flesh, but are partially arrested by bone, while others pass with almost equal facility through bone and flesh.

It seems to me that in these rays we may have a possible mode of transmitting intelligence which, with a few reasonable postulates, may supply a key to much that is obscure in psychical research. Let it be assumed that these rays, or rays even of higher frequency, can pass into the brain and act on some nervous center there. Let it be conceived that the brain contains a center which uses these rays as the vocal chords use sound vibrations (both being under the command of intelligence), and sends them out, with the velocity of light, to impinge on the receiving ganglion of another brain. In this way some, at least, of the phenomena of telepathy, and the transmission of intelligence from one sensitive to another through long distances, seem to come into the domain of law and can be grasped. A sensitive may be one who possesses the telepathic transmitting or receiving ganglion in an advanced state of development, or who, by constant practice, is rendered more sensitive to these high-frequency waves. Experience seems to show that the receiving and the transmitting ganglions are not equally developed; one may be active, while the other, like the pineal eye in man, may be only vestigial. By such an hypothesis no physical laws are violated; neither is it necessary to invoke what is commonly called the supernatural.

To this hypothesis it may be objected that brain waves, like any other waves, must obey physical laws. Therefore, transmission of thought must be easier or more certain the nearer the agent and recipient are to each other, and should die out altogether before great distances are reached. Also it can be urged that if brain waves diffuse in all directions they should affect all sensitives within their radius of action, instead of impressing only one brain. The electric telegraph is not a parallel case, for there a material wire intervenes to conduct and guide the energy to its destination.

CHAPTER 55

ERNST HAECKEL

ERNST HEINRICH PHILIPP AUGUST Haeckel was born February 16, 1834, in Potsdam, Prussia. He received a thorough education and later became Professor of Zoology at Jena. We might refer to him as the Darwin of his country. He expounded the theory that there is a definite relationship between the evolution and development of the embryo and the development of its species. We can trace back through the embryo, he declared, the development of a race. It is now definitely known that the human embryo, in its process of development, goes through various stages which correspond to the embryonic forms of the lesser animals before it attains that of man. Haeckel's investigations on the lower classes of marine organisms are also classical.

It is with pleasure, therefore, that we bring to you below excerpts from his famous thesis entitled, *The Fundamental Law of the Evolution of Organisms*. We advise every student, whether particularly interested in biology and zoology or not, to carefully read the excerpts below. They will find them provocative of thought. The more complete works of Haeckel may be obtained in any large public library.

THE FUNDAMENTAL LAW OF THE EVOLUTION OF ORGANISMS

The natural phenomena of the evolutionary history of man claim an entirely peculiar place in the wide range of the scientific study of nature. There is surely no subject of scientific investigation touching man more closely, or in the knowledge of which he is more deeply concerned, than the human organism itself; and of all the various branches of the science of man, or anthropology, the history of his natural evolution should excite his highest interest. For it affords a key for the solution of the greatest of those problems with which human science is occupied—the inquiry into the true nature of man, or, as it is called, the question of "Man's Place in Nature," which deals with the past and primitive history, the present condition, and future of Man—are all most directly and intimately linked to this branch of scientific research, which is called

The History of the Evolution of Man, or briefly, "Anthropogeny." It is however, a most astonishing but incontestable fact, that the history of the evolution of man as yet constitutes no part of general education. Indeed, our so-called "educated classes" are to this day in total ignorance of the most important circumstances and the most remarkable phenomena which Anthropogeny has brought to light.

In corroboration of this most astounding fact, I will only mention that most "educated people" do not even know that each human individual is developed from an egg, and that this egg is a simple cell, like that of any animal or plant. They are also ignorant of the fact that, in the development of this egg, an organism is first formed which is entirely different from the fully developed human body, to which it bears no trace of resemblance. The majority of "educated people" have never seen such a human germ, or embryo, in the early stages of development, nor are they aware that it is not at all different from those of other animals. They do not know that, at a certain period, this embryo has essentially the anatomical structure of a lancelet, later of a fish, and in subsequent stages those of amphibian and mammal forms; and that in the further evolution of these mammal forms those first appear which stand lowest in the series, namely, forms allied to the beaked animals *(Ornithorhynchus)*; then those allied to pouched animals *(Marsupialia)*, which are followed by forms most resembling apes; till at last the peculiar human form is produced as the final result. These significant facts are so little known that, when incidentally mentioned, they are commonly doubted, or are even regarded as unfounded inventions. Every one knows that the butterfly proceeds from a pupa, the pupa from a caterpillar, to which it bears no resemblance, and again the caterpillar from the egg of the butterfly. But few, except those of the medical profession, are aware that man, in the course of his individual evolution, passes through a series of transformations no less astonishing and remarkable than the well-known metamorphoses of the butterfly. The mere tracing of this wonderful series of forms, through which the human embryo passes in the course of its development, is, of course, of great general interest. But our understanding will be satisfied in a far higher degree, if we refer these remarkable facts to their final causes, and recognize that these natural phenomena are of the utmost importance to the entire range of human knowledge. They are of special importance to the "History of Creation," and, in connection with this, to philosophy in general—as we shall presently see. Further, as the general results of all human striving after knowledge are summed up in philosophy, it follows that every branch of scientific research comes more or less in contact with, and is influenced by, the History of the Evolution of Man.

In undertaking to describe the most important characteristics of these significant phenomena, and to trace them back to their final causes, I shall assign a much greater scope and aim to the History of the Evolution of Man than is usual. The lectures given on this subject in German universities during the past fifty years have been exclusively designed for medical students. It is true that the physician is most deeply interested in becoming acquainted with the development of the bodily organization of man, with which he deals, practically, from day to day, in his profession. I shall not here attempt to give a special account of the course of the evolution of the individual, such as has usually been given in embryological lectures, because few of my readers have studied human anatomy, or are acquainted with the physical structure of the developed man. Hence, I shall have to confine myself in many points to general outlines, neglecting many of the remarkable details, which would have to be discussed in treating of the evolution of special human organs, but which from their complicated nature, and because they are not easy to describe, can only be completely understood by the aid of an intimate acquaintance with human anatomy. I shall strive, however, to present this branch of the science in as popular a form as possible. A satisfactory general idea of the course of the evolution of the human embryo can, indeed, be given without going very deeply into anatomical details. As numerous successful attempts have recently been made to awaken the interest of larger classes of educated persons in other branches of science, I also may hope to succeed in this department, though it is in many respects especially beset with difficulties.

The History of the Evolution of Man, as it has been usually treated in lectures for medical students at the universities, has only concerned itself with Embryology, so-called, or more correctly with Ontogeny; in other words, with the history of the evolution of individual human organisms. This, however, is only the first part of the task before us, only the first half of the History of the Evolution of Man in the wider sense which will here be attributed to the term. The second part, equal in importance and interest, is Phylogeny, which is the history of the evolution of the descent of man; that is, of the evolution of the various animal forms through which, in the course of countless ages, mankind has gradually passed into its present form. All my readers know of the very important scientific movement which Charles Darwin caused fifteen years ago by his book on the Origin of Species. The most important direct consequence of this work, which marks a fresh epoch, has been to cause new inquiries to be made into the origin of the human race, which have proved the natural evolution of man through lower animal forms. The science which treats of the development of the human race from the animal kingdom is called Phylogeny, or the tribal history of man. The most important source from which the science derives its material is Ontogeny, or the history of germs; in other words, of the evolution of the individual. Paleontology, or the science of petrifactions, and, in a yet greater degree, comparative anatomy, also afford most important aid to Phylogeny.

These two divisions of our science, Ontogeny, or the history of the germ, Phylogeny, or the history of the tribe, are the most intimately connected, and the one cannot be understood without the other. The close intertwining of both branches, the increased proportions which germ-history and tribal history lend to each other, alone raise Biogeny (or the history of organic evolution, in the widest sense) to the rank of a philosophic natural science. The connection between the two is not external and superficial, but deeply internal and casual. Our knowledge of this connection has been but very recently obtained; it is most clearly and accurately expressed in the comprehensive statement which I call "the fundamental law of organic evolution," or more briefly, "the first principle of Biogeny."

This fundamental law, to which we shall recur again and again, and on the recognition of which depends the thorough understanding of the history of evolution, is briefly expressed in the proposition: that the History of the Germ is an epitome of the History of the Descent; or, in other words; that Ontogeny is a recapitulation of Phylogeny; or, somewhat more explicitly: that the series of forms through which the individual organism passes during its progress from the egg cell to its fully developed state, is a brief, compressed reproduction of the long series of forms through which the animal ancestors of that organism (or the ancestral forms of its species) have passed from the earliest periods of so-called organic creation down to the present time.

The casual nature of the relation which connects the History of the Germ (Embryology, or Ontogeny) with that of the tribe (Phylogeny) is dependent on the phenomena of heredity and adaptation. When these are properly understood, and their fundamental importance in determining the forms of organisms recognized, we may go a step further, and say: Phylogenesis is the mechanical cause of Ontogenesis. The Evolution of the Tribe, which is dependent on the laws of heredity and adaptation, affects all the events which take place in the course of the Evolution of the Germ or Embryo.

CHAPTER 56

JOHN FISKE

JOHN FISKE, THE AMERICAN historian and philosopher, was born in Hartford, Connecticut, on March 30, 1842. He did not begin college very early in life and before entering he was widely read in English literature, history, and ancient and modern languages. After graduating at Harvard he continued to study languages and philosophy. He had a great love for the latter subject.

He prepared himself for law—went two years to Harvard Law School, and upon his graduation opened an office in Boston. But most of his time was devoted to writing for various periodicals. In 1869 he gave a course of lectures at Harvard on the positive philosophy and delivered some thirty-five lectures on the doctrine of evolution, which he afterward expanded and which became the *Outlines of Cosmic Philosophy*. Because of the lucidity of his style, he contributed much to America's knowledge of Darwin and Spencer.

Perhaps his greatest contribution in the field of literature was his demonstration that religion and the doctrine of evolution were not, as believed, incompatible. He died in Gloucester, Massachusetts, on July 4, 1901. He spent the majority of his life in Cambridge.

We find that his philosophy follows mainly along mystical and metaphysical lines—that is, although his subject or topic may be of a physical nature, the trend is toward the mystical.

COSMIC PHILOSOPHY

As regards the significance of Man's position in the universe, this gradual elimination of strife is a fact of utterly unparalleled grandeur. Words cannot do justice to such a fact. It means that the wholesale destruction of life, which has heretofore characterized evolution ever since life began, and through which the higher forms of organic existence have been produced, must presently come to an end in the case of the chief of God's creatures. It means that the universal struggle for existence, having succeeded in bringing forth that consummate product of creative energy, the Human Soul, has done its work and will presently ease. In the lower regions of organic life it must go on, but as a determining factor in the highest work of evolution it will disappear.

The action of natural selection upon Man has long since been essentially diminished through the operation of social conditions. For in all grades of civilization above the lowest, "there are so many kinds of superiorities which severally enable men to survive, notwithstanding accompanying inferiorities, that natural selection cannot by itself rectify any particular unfitness." In a race of inferior animals any maladjustment is quickly removed by natural selection, because, owing to the universal slaughter, the highest completeness of life possible to a given grade of organization is required for the mere maintenance of life. But under the conditions surrounding human development it is otherwise. There is a wide interval between the highest and lowest degrees of completeness of living that are compatible with maintenance of life. Hence the wicked flourish. Vice is but slowly eliminated, because mankind has so many other qualities, beside the bad ones, which enable it to subsist and achieve progress in spite of them, that natural selection—which always works through death—cannot come into play. The improvement of civilized man goes on mainly through processes of direct adaptation. The principle in accordance with which the gloved hand of the dandy becomes white and soft while the hand of the labouring man grows brown and tough is the main principle at work in the improvement of Humanity. Our intellectual faculties, our passions and prejudices, our tastes and habits, become strengthened by use and weakened by disuse, just as the blacksmith's arm grows strong and the horse turned out to pasture becomes unfit for work. This law of use and disuse has been of immense importance throughout the whole evolution of organic life. With Man it has come to be paramount.

If now we contrast the civilized man intellectually and morally with the savage, we find that, along with his vast increase of cerebral surface, he has an immensely greater power of representing in imagination objects and relations not present to the senses. This is the fundamental intellectual difference between civilized men and savages. The power of imagination, or ideal representation, underlies the whole of science and art, and it is closely connected with the ability to work hard and submit to present discomfort for the sake of a distant reward. It is also closely connected with the development of the sympathetic feelings. The better we can imagine objects and relations not present to sense, the more readily we can sympathize with other people. Half the cruelty in the world is the direct result of stupid incapacity to put one's self in the other man's place. So closely interrelated are our intellectual and moral natures that the development of sympathy is very considerably determined by increasing width and variety of experience. From the simplest form of sympathy, such as the painful thrill felt on seeing some one in a dangerous position, up to the elaborate complication of altruistic feelings involved in the notion of abstract justice, the development is very largely a development of the representative faculty. The very same causes, therefore, deeply grounded in the nature of industrial civilization, which have developed science and art, have also had a distinct tendency to encourage the growth of the sympathetic emotions.

But, as already observed, these emotions are still too feebly developed, even in the highest races of men. We have made more progress in intelligence than in kindness. For thousands of generations, and until very recent times, one of the chief occupations of men has been to plunder, bruise, and kill one another. The selfish and ugly passions which are primordial—which have the incalculable strength of inheritance from the time when animal consciousness began—have had but little opportunity to grow weak from disuse. The tender and unselfish feelings, which are a later product of evolution, have too seldom been allowed to grow strong from exercise. And the whims and prejudices of the primeval militant barbarism are slow in dying out from the midst of peaceful industrial civilization. The coarser forms of cruelty are disappearing, and the butchery of men has greatly diminished. But most people apply to industrial pursuits a notion of antagonism derived from ages of warfare, and seek in all manner of ways to cheat or overreach one another. And as in more barbarous times the hero was he who had slain his tens of thousands, so now the man who has made wealth by overreaching his neighbours is not uncommonly spoken of in terms which imply approval. Though gentlemen, moreover, no longer assail one another with knives and clubs, they still inflict wounds with cruel words and sneers. Though the freethinker is no longer chained to a stake and burned, people still tell lies about him, and do their best to starve him by hurting his reputation. The virtues of forbearance and self-control are still in a very rudimentary state, and of mutual helpfulness there is far too little among men.

Nevertheless in all these respects some improvement has been made, along with the diminution of warfare, and by the time warfare has not merely ceased from the earth but has come to be the dimly remembered phantom of a remote past, the development of the sympathetic side of human nature will doubtless become prodigious. The manifestation of selfish and hateful feelings will be more and more sternly repressed by public opinion, and such feelings will become weakened by disuse, while the sympathetic feelings will increase in strength as the sphere for their exercise is enlarged. As thus at length we see what human progress means. It means throwing off the brute-inheritance,—gradually throwing it off through ages of struggle that are by and by to make struggle needless. Man is slowly passing from a primitive social state in which he was little better than a brute, toward an ultimate social state in which his character shall have become so transformed that nothing of the brute can be detected in it. The ape and the tiger in human nature will become extinct. Theology has had much to say about original sin. This original sin is neither more nor less than the brute-inheritance which every man carries with him, and the process of evolution is an advance toward true salvation. Fresh value is thus added to human life. The modern prophet, employing the methods of science, may again proclaim that the kingdom of heaven is at hand. Work ye, therefore, early and late, to prepare its coming.

The Message of Christianity

Now what is this message of the modern prophet but pure Christianity?—not the mass of theological doctrine ingeniously piled up by Justin Martyr and Tertullian and Clement and Athanasius and Augustine, but the real and essential Christianity which came, fraught with good tidings to men, from the very lips of Jesus and Paul! When did St. Paul's conception of the two men within him that warred against each other, the appetites of our brute nature and the God-given yearning for a higher life,—when did this grand conception ever have so much significance as now? When have we ever before held such a clew to the meaning of Christ in the Sermon on the Mount? "Blessed are the meek, for they shall inherit the earth." In the cruel strife of centuries has it not often seemed as if the earth were to be rather the prize of the hardest heart and the strongest fist? To many men these words of Christ have been as foolishness and as a stumbling-block, and the ethics of the Sermon on the Mount have been openly derided as too good for this world. In that wonderful picture of modern life which is the greatest work of one of the great seers of our time, Victor Hugo gives a concrete illustration of the working of Christ's methods. In the saint-like career of Bishop Myriel, and in the transformation which his example works in the character of the hardened outlaw Jean Valjean, we have a most powerful commentary on the Sermon on the Mount. By some critics who could express their views freely about "Les Miserables" while hesitating to impugn directly the authority of the New Testament, Monseigneur Bienvenu was unsparingly ridiculed as a man of impossible goodness, and as a milksop and fool withal. But I think Victor Hugo understood the capabilities of human nature, and its real dignity, much better than these scoffers. In a low stage of civilization Monseigneur Bienvenu would have had small chance of reaching middle life. Christ himself, we remember, was crucified between two thieves. It is none the less true that when once the degree of civilization is such as to allow this highest type of character, distinguished by its meekness and kindness, to take root and thrive, its methods are incomparable in their potency. The Master knew full well that the time was not yet ripe,—that he brought not peace, but sword. But he preached nevertheless that gospel of great joy which is by and by to be realized by toiling Humanity, and he announced ethical principles fit for the time that is coming. The great originality of his teaching, and the feature that has chiefly given it power in the world, lay in the distinctness with which he conceived a state of society from which every vestige of strife, and the modes of behaviour adapted to ages of strife, shall be utterly and forever swept away. Through misery that has seemed unendurable and turmoil that has seemed endless, men have thought on that gracious life and its sublime ideal, and have taken comfort in the sweetly solemn message of peace on earth and good will to men.

I believe that the promise with which I started has now been amply redeemed. I believe it has been fully shown that so far from degrading Humanity, or putting it on a level with the animal world in general, the doctrine of evolution shows us distinctly for the first time how the creation and the perfecting of Man is the goal toward which Nature's work has been tending from the first. We can now see clearly that our new knowledge enlarges tenfold the significance of human life, and makes it seem more than ever the chief object of Divine care, the consummate fruition of that creative energy which is manifested throughout the knowable universe.

CHAPTER 57

ELBERT HUBBARD

ELBERT GREEN HUBBARD, EMINENT American philosopher and essayist, was born in Bloomington, Illinois, in 1856. His education in his youth was meager. He had a common school education and supported himself by working on a farm and in a printing office. Later, he devoted himself to private study and extensive travel. He concluded his travels by settling at East Aurora, New York, where he established the renowned Roycroft Press.

The magnificent typography of the periodicals and publications disseminated from that establishment made them recognized examples of the highest in the printing art. His first publication was a little magazine known as *The Philistine* which almost immediately attracted attention. An essay appearing in one of its issues entitled, "A Message to Garcia," was so well received that it was later published in pamphlet form and eventually had a total circulation throughout the world of fifteen million copies.

When about forty years of age, he entered Harvard College and for a period of three years studied literature and language. Mr. Hubbard was not only a forceful thinker and writer, but brought additional laurels to himself by his eloquent manner of speaking and his splendid platform appearance. He was especially inclined toward mysticism and occult and metaphysical studies and researches. He sponsored numerous small societies in their investigations of the mysteries of nature. He was intimately known to Dr. H. Spencer Lewis at the time of the formation of the present jurisdiction of the Rosicrucian Order in North and South America. He died in 1915.

Below are a few of his numerous essays. Every reader will be impressed with the simplicity of his style, its beauty, and the fact that each thought expressed is exceedingly cogent.

THE TEACHER

It is a great thing to teach. To give yourself in a way to inspire others to think, to do, to become—what nobler ambition! To be a good teacher demands a high degree of altruism, for one must be willing to sink self, to die, as it were, that others may live. There is somthing in it that is akin to motherhood—a brooding quality. Every true mother realizes that her children are only loaned to her—sent from God—and the attributes of her mind and body are being used by some Power for a Purpose. The teacher is training her children to do without her.

Desire

What is it wins? Work you say, but you are wrong. It is desire that brings every good thing. Did you ever watch a cat about to spring for a bird? The cat does not think about working to secure that bird: about how to place its body for the most graceful spring—not that. It is just filled with the desire, and it does exactly the proper thing—the single-hearted thing. Rabbits can run faster and farther than cats, but rabbits never catch birds—they do not desire to.

Eternity

We are living in eternity now, just as much as we ever shall. God is right here now, and we are as near Him now as we shall ever be. He never started this world a-going and went away and left it—He is with us yet. There is no devil but fear, and nobody and nothing can harm you but yourself. We should remember the weekday to keep it holy, live one day at a time, doing our work the best we can. There is no more sacred place than that where a man is doing good and useful work, and there is no higher wisdom than to lose yourself in useful industry, and be kind—and be kind.

My Creed

I wish to be simple, honest, natural, frank, clean in mind and clean in body, unaffected—ready to say, "I do not know," if so it be, to meet all men on an absolute equality—to face any obstacle and meet every difficulty unafraid and unabashed. I wish to live without hate, whim, jealousy, envy or fear. I wish others to live their lives, too—up to their highest, fullest and best. To that end I pray that I may never meddle, dictate, interfere, give advice that is not wanted, nor assist when my services are not needed. If I can help people, I will do it by giving them a chance to help themselves; and if I can uplift or inspire, let it be by example, inference and suggestion, rather than by injunction and dictation. I desire to Radiate Life!

Fear and Doubt

The world accepts a man at the estimate he places upon himself. Many men are strong at times, but strong men make enemies—they have detractors—calumny calls and hate hisses. Then doubt comes creeping in, possibly the enemies are right—ah, who knows! And instantly the doubt is communicated to the public—the man's face tells his fears to all he meets. And their estimate of the man is the lowest standard he sets upon himself.

That is why we need Some One to believe in us—if we do well, we want our work commended, our faith corroborated.

So note this, when you find the strong man he is one who is well sustained.

To associate closely with those who doubt or distrust you is eventually going to make you distrust yourself. And then we get dead conformity, hopeless mediocrity, nothing more. The individual who thinks well of you, who keeps his mind on your good qualities, and does not look for flaws, is your friend. Who is my brother? I'll tell you, he is one who recognizes the good in me.

On Walt Whitman

Most writers bear no message—they carry no torch. Sometimes they excite wonder, or they amuse and divert—divert us from our work. To be diverted to a certain degree may be well, but there is a point where earth ends and cloudland begins, and even great poets occasionally befog the things which they would reveal.

Homer was seemingly blind to much simple truth; Virgil carries you away from earth; Horace was undone without his Macaenas; Dante makes you an exile; Shakespeare was singularly silent concerning the doubts, difficulties, and common lives of common people; Byron's Corsair life does not help you in your toil, and in his fight with English Bards and Scotch Reviewers we crave neutrality; to be caught in the meshes of Pope's *Dunciad* is not pleasant; and Lowell's *Fable for Critics* is only another *Dunciad*. But above all poets who have ever lived, the author of *Leaves of Grass* was the poet of humanity.

Milton knew all about Heaven, and Dante conducts us through Hell, but it was left for Whitman to show us Earth. His voice never goes so high that it breaks an impotent falsetto, neither does it growl and snarl at things it does not understand, and, not understanding, does not like. He was so great that he had no envy, and his insight was so sure that he had no prejudice. He never boasted that he was higher, nor claimed to be less than any of the other sons of men. He met all on terms of absolute equality, mixing with the poor, the lowly, the fallen, the oppressed, the cultured, the rich—

simply as brother with brother. And when he said to the outcast, "Not till the sun excludes you will I exclude you," he voiced a sentiment worthy of a god.

He was brother to the elements, the mountains, the seas, the clouds, the sky. He loved them all and partook of them all in his large, free, unselfish, untrammeled nature. His heart knew no limits, and feeling his feet mortis'd in granite and his footsteps tenon'd in infinity, he knew the amplitude of time.

Only the great are generous; only the strong are forgiving. Like Lot's wife, most poets look back over their shoulders; and those who are not looking backward insist that we shall look into the future, and the vast majority of the whole scribbling rabble accept the precept, "Man never is, but always to be blest."

We grieve for childhood's happy days, and long for sweet rest in Heaven, and sigh for mansions in the skies. And the people about us seem so indifferent, and our friends so lukewarm; and really no one understands us, and our environment queers our budding spirituality and the frost of jealousy nips our aspirations: "Oh Paradise, oh Paradise, the world is growing old; who would not be at rest and free where love is never cold." So sing the fearsome dyspeptics of the stylus. Oh anemic he, you bloodless she, nipping at crackers, sipping at tea, why not consider that although the evolutionists tell us where we came from, and the theologians inform us where we are going to, yet the only thing we are really sure of is that we are here!

The present is the perpetually moving spot where history ends and prophecy begins. It is our only possession—the past we reach through lapsing memory, halting recollection, hearsay, and belief; we pierce the future by wistful faith or anxious hope, but the present is beneath our feet.

Whitman sings the beauty and the glory of the present. He rebukes our groans and sighs—bids us look about on every side at the wonders of creation, and at the miracles within our grasp. He lifts us up, restores us to our own, introduces us to man and Nature and thus infuses into us courage, manly pride, self-reliance, and the strong faith that comes when we feel our kinship with God.

He was so mixed with the universe that his voice took in the sway of elemental integrity and candor. Absolutely honest, this man was unafraid and unashamed, for Nature has neither apprehension, shame nor vain-glory. In *Leaves of Grass* Whitman speaks as all men have ever spoken who believe in God and in themselves—oracular, without apology, without abasement—fearlessly. He tells of the powers and mysteries that pervade and guide all life, all death, all purpose. His work is masculine, as the sun is masculine; for the Prophetic voice is as surely masculine as the lullaby and lyric cry is feminine.

Whitman brings the warmth of the sun to the buds of the heart so that they open and bring forth form, color, perfume. He becomes for them aliment and dew; so these buds become blossoms, fruits, tall branches, and stately trees that cast refreshing shadows.

There are men who are to other men as the shadow of a mighty rock in a weary land—such is Walt Whitman.

CHAPTER 58

ALBERT EINSTEIN

ALBERT EINSTEIN WAS A German-Swiss, and was born of Jewish parents May 14, 1879, in Ulm, Württemberg. His boyhood was spent at Munich, where his father owned an electro-technical works. The family moved to Italy in 1894. Albert Einstein at that time went to Cantonal School at Aarau in Switzerland. He later attended lectures while supporting himself by teaching mathematics and physics at Polytechnic school in Zurich until 1900. He was given the position of examiner of patents in the patent office at Berne. He acquired the much-coveted Ph.D. at the University of Zurich and published his first paper on physical matters. In 1911, he accepted the chair of physics in Prague. In 1913 began the period of his prominence. His reputation as a writer on abstract subjects spread. He was appointed director of the Kaiser-Wilhelm Physical Institute in Berlin. Later he was made a member of the Royal Prussian Academy of Sciences. Honorary degrees, as his fame spread, were conferred upon him by the Universities of Amsterdam, Copenhagen, Geneva, Manchester, Rostock, and Princeton. He died in 1955.

His writings were not confined to abstraction. Many famous practical papers on heat, temperature, and molecular physics were prepared by him. Following, we bring to you excerpts from his famous Space-Time Theory based, of course, upon his Theory of Relativity. Broadly speaking, the uniqueness of his Space-Time Theory is that he makes space and time absolutely dependent upon each other, which before our era were not so considered. Even during a fairly recent period of science and advanced mathematics there had not been recognized a relationship or dependency between time and space.

The excerpts from the following article are worthy of your most careful study and consideration. If it is possible to sum them up briefly, we can say that they convey the idea that consciousness confers identity upon our sense experiences. Time measures the duration of our consciousness of these sense experiences, and space, or the three dimensions, is the measurement of the relative area of one object of our sense experience with that of another. We find, therefore, that not only are space and time closely related to each other and dependent upon each other, but also consciousness, and we can therefore refer to the subject as the Consciousness-Space-Time-Theory.

SPACE-TIME

The Theory of Relativity has brought about a fundamental change in the scientific conception of space and time, described in a famous saying of Minkowski—"From henceforth space in itself and time in itself sink to mere shadows, and only a kind of union of the two preserves an independent existence."

All our thoughts and concepts are called up by sense-experiences and have a meaning only in reference to these sense-experiences. On the other hand, however, they are products of the spontaneous activity of our minds; they are thus in no wise logical consequences of the contents of these sense-experiences. If, therefore, we wish to grasp the essence of a complex of abstract notions we must for the one part investigate the mutual relationships between the concepts and the assertions made about them; for the other, we must investigate how they are related to the experiences.

So far as the way is concerned in which concepts are connected with one another and with the experiences there is no difference of principle between the concept-systems of science and those of daily life. The concept-systems of science have grown out of those of daily life and have been modified and completed according to the objects and purposes of the science in question.

The more universal a concept is the more frequently it enters into our thinking; and the more indirect its relation to sense-experience, the more difficult it is for us to comprehend its meaning; this is particularly the case with prescientific concepts that we have been accustomed to use since childhood. Consider the concepts referred to in the words "where," "when," "why," "being," to the elucidation of which innumerable volumes of philosophy have been devoted. We fare no better in our speculations than a fish which should strive to become clear as to what is water.

Space

In the present article we are concerned with the meaning of "where," that is, of space. It appears that there is no quality contained in our individual primitive sense-experiences that may be designated as spatial. Rather, what is spatial appears to be a sort of order of the material objects of experience. The concept "material object" must therefore be available if concepts concerning space are to be possible. It is the logically primary concept. This is easily seen if we analyze the spatial concepts for example, "next to," "touch," and so forth, that is, if we strive to become aware of their equivalents in experience. The concept "object" is a means of taking into account the persistence in time or the continuity, respectively, of certain groups of experience-complexes. The existence of objects is thus of a conceptual nature, and the meaning of the concepts of objects depends wholly on their being connected (intuitively) with groups of elementary sense-experiences. This connection is the basis of the illusion which makes primitive experience appear to inform us directly about the relation of material bodies (which exist, after all, only in so far as they are thought).

In the sense thus indicated we have (the direct) experience of the contact of two bodies. We need do no more than call attention to this, as we gain nothing for our present purpose by singling out the individual experiences to which this assertion alludes. Many bodies can be brought into permanent contact with one another in manifold ways. We speak in this sense of the position-relationships of bodies (Lagenbeziehungen). The general laws of such position-relationships are essentially the concern of geometry. This holds, at least, if we do not wish to restrict ourselves to regarding the propositions that occur in this branch of knowledge merely as relationships between empty words that have been set up according to certain principles.

Pre-Scientific Thought

Now, what is the meaning of the concept "space" which we also encounter in pre-scientific thought? The concept of space in pre-scientific thought is characterized by the sentence: "we can think away things but not the space which they occupy." It is as if, without having had experience of any sort, we had a concept, nay even a presentation, of space and as if we ordered our sense-experiences with the help of this concept, present *a priori*. On the other hand, space appears as a physical reality, as a thing which exists independently of our thought, like material objects. Under the influence of this view of space the fundamental concepts of geometry: the point, the straight line, the plane, were even regarded as having a self-evident character. The fundamental principles that deal with these configurations were regarded as being necessarily valid and as having at the same time an objective content. No scruples were felt about ascribing an objective meaning to such statements as "three empirically given bodies (practically infinitely small) lie on one straight line," without demanding a physical definition for such an assertion. This blind faith in evidence and in the immediately real meaning of the concepts and propositions of geometry became uncertain only after non-Euclidean geometry had been introduced.

Reference to the Earth

If we start from the view that all spatial concepts are related to contact-experiences of solid bodies, it is easy to understand how the concept "space" originated, namely, how a thing independent of bodies and yet embodying their position-possibilities (Lagerungsmoglichkeiten) was posited. If we have a system of bodies in contact and at rest relatively to one another, some can be replaced by others. This property of allowing substitution is interpreted as "available space." Space denotes the property in virtue of which rigid bodies can occupy different positions. The view that space is something with a unity of its own is perhaps due to the circumstance that in pre-scientific thought all positions of bodies were referred to one body (reference body), namely the earth. In scientific thought the earth is represented by the co-ordinate system. The assertion that it would be possible to place an unlimited number of bodies next to one another denotes that space is infinite. In pre-scientific thought the concepts "space" and "time" and "body of reference" are scarcely differentiated at all. A place or point in space is always taken to mean a material point on a body of reference.

Time

The physical time-concept answers to the time-concept of the extra-scientific mind. Now, the latter has its root in the time-order of the experiences of the individual, and this order we must accept as something primarily given. One experiences the moment "now," or, expressed more accurately, the present sense-experience *(Sinnen-Erlebnis)* combined with the recollection of (earlier) sense-experiences. That is why the sense-experiences seem to form a series, namely the time-series indicated by "earlier" and "later." The experience series is thought of as a one-dimensional continuum. Experience-series can repeat themselves and can then be recognized. They can also be repeated inexactly, wherein some events are replaced by others without the character of the repetition becoming lost for us. In this way we form the time-concept as a one-dimensional frame which can be filled in by experiences in various ways. The same series of experiences answer to the same subjective time-intervals.

The transition from this "subjective" time *(Ich-Zeit)* to the time-concept of pre-scientific thought is connected with the formation of the idea that there is a real external world independent of the subject. In this sense the (objective) event is made to correspond with the subjective experience. In the same sense there is attributed to the "subjective" time of the experience a "time" of the corresponding "objective" event. In contrast with experiences, external events and their order in time claim validity for all subjects.

This process of objectification would encounter no difficulties were the time-order of the experiences corresponding to a series of external events the same for all individuals. In the case of the immediate visual perceptions of our daily lives, this correspondence is exact. That is why the idea that there is an objective time-order became established to an extraordinary extent. In working out the idea of an objective world of external events in greater detail, it was found necessary to make events and experiences depend on each other in a more complicated way. This was at first done by means of rules and modes of thought instinctively gained, in which the conception of space plays a particularly prominent part. This process of refinement leads ultimately to natural science.

The measurement of time is effected by means of clocks. A clock is a thing which automatically passes in succession through a (practically) equal series of events (period). The number of periods (clock-time) elapsed serves as a measure of time. The meaning of this definition is at once clear if the event occurs in the immediate vicinity of the clock in space; for all observers then observe the same clock-time simultaneously with the event (by means of the eye) independently of their position. Until the theory of relativity was propounded it was assumed that the conception of simultaneity had an absolute objective meaning also for events separated in space.

This assumption was demolished by the discovery of the law of propagation of light. For if the velocity of light in empty space is to be a quantity that is independent of the choice (or, respectively, of the state of motion) of the inertial system to which it is referred, no absolute meaning can be assigned to the conception of the simultaneity of events that occur at points separated by a distance in space. Rather, a special time must be allocated to every inertial system. If no co-ordinate system (inertial system) is used as a basis of reference there is no sense in asserting that events at different points in space occur simultaneously. It is in consequence of this that space and time are welded together into a uniform four-dimensional continuum.

Explanatory

THE ROSICRUCIAN ORDER

Purpose and Work of the Order

Anticipating questions which may be asked by the readers of this book, the publishers take this opportunity to explain the purpose of this Order and how you may learn more about it.

There is only one universal Rosicrucian Order existing in the world today, united in its various jurisdictions, and having one Supreme Council in accordance with the original plan of the ancient Rosicrucian manifestoes. The Rosicrucian Order is not a religious or sectarian society.

This international organization retains the ancient traditions, teachings, principles, and practical helpfulness of the Order as founded centuries ago. It is known as the *Ancient Mystical Order Rosae Crucis,* which name, for popular use, is abbreviated into AMORC. The Headquarters of the Worldwide Jurisdiction (The Americas, Australasia, Europe, Africa, and Asia) is located at San Jose, California.

The Order is primarily a humanitarian movement, making for greater Health, Happiness, and Peace in people's *earthly lives*, for we are not concerned with any doctrine devoted to the interests of individuals living in an unknown, future state. The Work of Rosicrucians is to be done *here* and *now*; not that we have neither hope nor expectation of *another* life after this, but we *know* that the happiness of the future depends upon *what we do today for others* as well as for ourselves.

Secondly, our purposes are to enable men and women to live clean, normal, natural lives, as Nature intended, enjoying *all* the privileges of Nature, and all benefits and gifts equally with all of humanity; and to be *free* from the shackles of superstition, the limits of ignorance, and the sufferings of avoidable *Karma*.

The Work of the Order—using the word "work" in an official sense, consists of teaching, studying, and testing such Laws of God and Nature as make our Members Masters in the Holy Temple (the physical body), and Workers in the Divine Laboratory (Nature's domains). This is to enable our Members to render *more efficient help* to those who do not know, and who need or require help and assistance.

Therefore, the Order is a School, a College, a Fraternity, with a laboratory. The Members are students and workers. The graduates are unselfish servants of God to Humanity, efficiently educated, trained, and experienced, attuned with the mighty forces of the Cosmic or Divine Mind, and Masters of matter, space, and time. This makes them essentially Mystics, Adepts, and Magi—creators of their own destiny.

There are no other benefits or rights. All Members are pledged to give unselfish Service, without other hope or expectation of remuneration than to Evolve the Self and prepare for a *greater* Work.

The Rosicrucian Sanctum membership program offers a means of personal home study. Instructions are sent once a month in specially prepared weekly lectures and lessons, and contain a summary of the Rosicrucian principles with such a wealth of personal experiments, exercises, and tests as will make each Member highly proficient in the attainment of certain degrees of mastership. The lectures are under the direction of the Imperator's staff. These correspondence lessons and lectures comprise several Degrees. Each Degree has its own Initiation ritual, to be performed by the Member at home in his or her private home sanctum. Such rituals are not the elaborate rituals used in the Lodge Temples, but are simple and of practical benefit to the student.

If you are interested in knowing more of the history and present-day helpful offerings of the Rosicrucians, you may receive a *free* copy of the booklet entitled *The Mastery of Life*, by sending a request to:

Scribe I.M.P.
Rosicrucian Order, AMORC
Rosicrucian Park
San Jose, California 95191, U.S.A.

ROSICRUCIAN LIBRARY

COSMIC MISSION FULFILLED

by Ralph M. Lewis, F.R.C.

This illustrated biography of Harvey Spencer Lewis, Imperator of the Ancient Mystical Order Rosae Crucis, was written in response to the requests of thousands of members who sought the key to this mystic-philosopher's life mission of rekindling the ancient flame of *Wisdom* in the Western world. We view his triumphs and tribulations from the viewpoint of those who knew him best.

Recognize, like him, that the present is our *moment in Eternity*; in it we fulfill our mission.

MENTAL POISONING
THOUGHTS THAT ENSLAVE MINDS

by H. Spencer Lewis, F.R.C., Ph. D.

Must humanity remain at the mercy of evil influences created in the minds of the vicious? Do poisoned thoughts find innocent victims? Use the knowledge this book fearlessly presents as an antidote for such superstitions and their influences.

There is no need to remain helpless even though evil thoughts of envy, hate, and jealousy are aimed to destroy your self-confidence and peace of mind.

THE SYMBOLIC PROPHECY OF THE GREAT PYRAMID

by H. Spencer Lewis, F.R.C., Ph. D.

The world's greatest mystery and first wonder is the Great Pyramid. Its history, vast wisdom, and prophecies are all revealed in this beautifully bound and illustrated book. You will be amazed at the pyramid's scientific construction and at the secret knowledge of its mysterious builders.

THE MYSTICAL LIFE OF JESUS

by H. Spencer Lewis, F.R.C., Ph. D.

A full account of Jesus' life, containing the story of his activities in the periods not mentioned in the Gospel accounts, *reveals the real Jesus* at last.

This book required a visit to Palestine and Egypt to secure verification of the strange facts found in Rosicrucian records. Its revelations, predating the discovery of the Dead Sea Scrolls, show aspects of the Essenes unavailable elsewhere.

This volume contains many mystical symbols (fully explained), photographs, and an unusual portrait of Jesus.

SEPHER YEZIRAH—A BOOK ON CREATION or The Jewish Metaphysics of Remote Antiquity

by Dr. Isidor Kalisch, Translator

The ancient basis for Kabalistic thought is revealed in this outstanding metaphysical essay concerning all creation. It explains the secret name of Jehovah.

Containing both the Hebrew and English texts, its 61 pages have been photolithographed from the 1877 edition. As an added convenience to students of Kabala, it contains a glossary of the original Hebraic words and terms.

SELF MASTERY AND FATE WITH THE CYCLES OF LIFE

by H. Spencer Lewis, F.R.C., Ph. D.

This book demonstrates how to harmonize the self with the cyclic forces of each life.

Happiness, health, and prosperity are available for those who know the periods in their own life that enhance the success of varying activities. Eliminate "chance" and "luck," cast aside "fate," and replace these with self mastery. Complete with diagrams and lists of cycles.

ROSICRUCIAN PRINCIPLES FOR THE HOME AND BUSINESS

by H. Spencer Lewis, F.R.C., Ph. D.

This volume contains the practical application of Rosicrucian teachings to such problems as: ill health, common ailments, how to increase one's income or promote business propositions. It shows not only what to do, but what to avoid, in using metaphysical and mystical principles in starting and bringing into realization new plans and ideas.

Both business organizations and business authorities have endorsed this book.

THE TECHNIQUE OF THE MASTER
or The Way of Cosmic Preparation

by Raymund Andrea, F.R.C.

A guide to inner unfoldment! The newest and simplest explanation for attaining the state of Cosmic Consciousness. To those who have felt the throb of a vital power within, and whose inner vision has at times glimpsed infinite peace and happiness, this book is offered. It converts the intangible whispers of self into forceful actions that bring real joys and accomplishments in life. It is a masterful work on psychic unfoldment.

ESSAYS OF A MODERN MYSTIC

by H. Spencer Lewis, F.R.C., Ph. D.

These private writings disclose the personal confidence and enlightenment that are born of *inner experience.* As a true mystic-philosopher, Dr. Lewis shares with his readers the results of contact with the Cosmic Intelligence residing within.

A THOUSAND YEARS OF YESTERDAYS

by H. Spencer Lewis, F.R.C., Ph.D.

This fascinating story dramatically presents the real facts of reincarnation. It explains how the soul leaves the body and *when* and *why* it returns to Earth again.

This revelation of the *mystic laws and principles* of the Masters of the East has never before been presented in such a form. Finely bound and stamped in gold, it makes a fine addition to your library.

EGYPT'S ANCIENT HERITAGE

by Rodman R. Clayson

Much of what we know today began in Egypt! Concepts that are unquestionably ancient show marvelous insights into natural law. From Egypt's mystery schools came rites and ceremonies that dramatize the creative Cosmic force. This book presents a masterly overview of the civilization of the Nile Valley.

MANSIONS OF THE SOUL

by H. Spencer Lewis, F.R.C., Ph. D.

Reincarnation—the world's most disputed doctrine! What did Jesus mean when he referred to the "mansions in my Father's house"? This book demonstrates what Jesus and his immediate followers knew about the rebirth of the soul, as well as what has been taught by sacred works and scholarly authorities in all parts of the world.

Learn about the cycles of the soul's reincarnations and how you can become acquainted with your present self and your past lives.

THE SECRET DOCTRINES OF JESUS

by H. Spencer Lewis, F.R.C., Ph. D.

Even though the sacred writings of the Bible have had their contents scrutinized, judged, and segments removed by twenty ecclesiastical councils since the year 328 A.D., there still remain buried in unexplained passages and parables the Great Master's *personal* doctrines.

Every thinking man and woman will find *hidden truths* in this book.

MYSTICS AT PRAYER

Compiled by Many Cihlar, F.R.C.

The first compilation of the famous prayers of the renowned mystics and adepts of all ages.

The book *Mystics at Prayer* explains in simple language the reason for prayer, how to pray, and the Cosmic laws involved. You come to learn the real efficacy of prayer and its full beauty dawns upon you. Whatever your religious beliefs, this book makes your prayers the application not of words, but of helpful, divine principles. You will learn the infinite power of prayer. Prayer is man's rightful heritage. It is the direct means of man's communion with the infinite force of divinity.

ETERNAL FRUITS OF KNOWLEDGE

by Cecil A. Poole, F.R.C.

A stimulating presentation of philosophical insights that will provoke you into considering new aspects of such questions as: the purpose of human existence, the value of mysticism, and the true nature of good and evil. Paperback.

THE TECHNIQUE OF THE DISCIPLE

by Raymund Andrea, F.R.C.

The Technique of the Disciple is a book containing a modern description of the ancient, esoteric path to spiritual Illumination, trod by the masters and avatars of yore. It has long been said that Christ left, as a great heritage to members of His secret council, a private method for guidance in life, which method has been preserved until today in the secret, occult, mystery schools.

Raymund Andrea, the author, reveals the method for attaining a greater life taught in these mystery schools, which perhaps parallels the secret instructions of Christ to members of his council. The book is informative, inspiring, and splendidly written. Paperback.

THE CONSCIOUS INTERLUDE

by Ralph M. Lewis, F.R.C.

With clarity of expression and insightful penetration of thought, this original philosopher leads us to contemplate such subjects as: the Fourth Dimension, the Mysteries of Time and Space; the Illusions of Law and Order; and many others of similar import.

As you follow the author through the pages into broad universal concepts, your mind too will feel its release into an expanding consciousness.

WHISPERINGS OF SELF

by Validivar

Wisdom, wit, and insight combine in these brief aphorisms that derive from the interpretation of Cosmic impulses received by Validivar, whose true name is Ralph M. Lewis, Imperator of the Rosicrucian Order.

These viewpoints of all areas of human experience make an attractive gift as well as a treasured possession of your own.

THE SANCTUARY OF SELF

by Ralph M. Lewis, F.R.C.

Are you living your life to your best advantage? Are you beset by a *conflict of desires*? Do you know that there are various *loves* and that some of them are dangerous drives?

Learn which of your feelings to discard as enslaving influences and which to retain as worthy incentives.

The author, Imperator of the Rosicrucian Order, brings to you from his years of experience, the practical aspects of mysticism.

ROSICRUCIAN QUESTIONS AND ANSWERS WITH COMPLETE HISTORY OF THE ORDER

by H. Spencer Lewis, F.R.C., Ph. D.

From ancient times to the present day, the history of the Rosicrucian Order is traced from its earliest traditional beginnings. Its historical facts are illuminated by stories of romance and mystery.

Hundreds of questions in this well-indexed volume are answered, dealing with the work, benefits, and purposes of the Order.

"UNTO THEE I GRANT..."

as revised by Sri Ramatherio

Out of the mysteries of the past comes this antique book that was written two thousand years ago, but was hidden in manuscript form from the eyes of the world and given only to the Initiates of the temples in Tibet to study privately.

It can be compared only with the writings attributed to Solomon in the Bible of today. It deals with man's passions, weaknesses, fortitudes, and hopes. Included is the story of the expedition into Tibet that secured the manuscript and the Grand Lama's permission to translate it.

IN SEARCH OF REALITY

by Cecil A. Poole, F.R.C.

This book unites metaphysics with mysticism. Man is not just an isolated entity on Earth. He is also of a great world—the Cosmos. The forces that create galaxies and island universes also flow through man's being. The human body and its vital phenomenon—Life—are of the same spectrum of energy of which all creation consists. The universe is you because you are one of its myriad forms of existence. Stripping away the mystery of this Cosmic relationship increases the personal reality of the Self. Paperback.

BEHOLD THE SIGN

by Ralph M. Lewis, F.R.C.

Unwrap the veil of mystery from the strange symbols inherited from antiquity. What were the *Sacred Traditions* said to be revealed to Moses? What were the discoveries of the Egyptian priesthood?

This book is fully illustrated with *age-old secret symbols* whose true meanings are often misunderstood. Even the mystical beginnings of the *secret signs* of many fraternal brotherhoods today are explained.

MYSTICISM—THE ULTIMATE EXPERIENCE

by Cecil A. Poole, F.R.C.

An experience is more than just a sensation, a feeling. It is an *awareness,* or perception, with *meaning.* Our experiences are infinite in number, yet they are limited to certain types. Some are related to our objective senses; others, to dreams and inspirational ideas. But there is *one* that transcends them all—the *mystical experience.* It serves every category of our being: it stimulates, it enlightens, it strengthens; it is the *Ultimate Experience.*

And this book, *Mysticism—The Ultimate Experience,* defines it in simple and inspiring terms.

LEMURIA—THE LOST CONTINENT OF THE PACIFIC

by Wishar S. Cervé

Where the Pacific now rolls in a majestic sweep for two thousand miles, there was once a vast continent known as Lemuria.

The scientific evidences of this lost race and its astounding civilization with the story of the descendants of the survivors present a cyclical viewpoint of rise and fall in the progress of civilization.

YESTERDAY HAS MUCH TO TELL

by Ralph M. Lewis, F.R.C.

A personal account of witnessing primitive ceremonies, conversations with mystical teachers and austere high priests of the Near and Far East. It takes you to rites in the interior of Africa and to temples in Peru, India, Egypt, and other exotic lands. This is no mere travel book; the author was privileged, because of his Rosicrucian affiliation, to see and learn that which is not ordinarily revealed.

SON OF THE SUN

by Savitri Devi

The amazing story of Akhnaton (Amenhotep IV), Pharaoh of Egypt, 1360 B.C. This is not just the fascinating story of one life—it is far more. It raises the curtain on man's emerging from superstition and idolatry. Against the tremendous opposition of a fanatical priesthood, Akhnaton brought about the world's first spiritual revolution. He was the first one to declare that there was a "sole God." In the words of Sir Flinders Petrie (*History of Egypt*): "Were it invented to satisfy our modern scientific conceptions, his religio-philosophy could not be logically improved upon at the present day."

This book contains over three hundred pages. It is handsomely printed, well bound, and stamped in gold.

ROSICRUCIAN MANUAL

by H. Spencer Lewis, F.R.C., Ph. D.

This practical book contains not only extracts from the Constitution of the Rosicrucian Order, but a complete outline and explanation of all the customs, habits, and terminology of the Rosicrucians, with diagrams and explanations of the symbols used in the teachings, an outline of the subjects taught, a dictionary of the terms, a complete presentation of the principles of Cosmic Consciousness, and biographical sketches of important individuals connected with the work. There are also special articles on the Great White Lodge and its existence, how to attain psychic illumination, the Rosicrucian Code of Life with twenty-nine laws and regulations, and a number of portraits of prominent mystics including Master K. H., the Illustrious.

The technical matter in the text and in the numerous diagrams makes this book a real encyclopedia of Rosicrucian explanations, aside from the dictionary of Rosicrucian terms.

The *Rosicrucian Manual* has been enlarged and improved since its first edition. Attractively bound and stamped in gold.

MESSAGES FROM THE CELESTIAL SANCTUM

by Raymond Bernard, F.R.C.

The real *unity* is Cosmic Unity. No human being is separated from the Cosmic, no matter where he lives or how different his life style may be. Each person is like a channel through which cosmically inspired intuitive impressions and guidance can flow. The *Celestial Sanctum* in general is the universe. No earthly sanctuary is more sacred than the multiple phenomena which occur in the great extensions of the Cosmic. There are no greater Laws than those which operate this phenomenon.

This book explains how you can harmonize yourself with the *Celestial Sanctum*. Also, it reveals rational, sensible, and practical messages which were *cosmically* narrated. They can guide all, regardless of race or creed, toward a greater understanding and a complete mastery of one's life. Allow this book to explain to you how your mind can become like a window through which you can observe creation—and learn from it in a *personal way*. Paperback.

HERBALISM THROUGH THE AGES

by Ralph Whiteside Kerr, F.R.C.

The seemingly magical power of herbs endowed them with a divine essence to the mind of early man. Not only did they provide some of his earliest foods and become medicines for his illnesses but they also symbolized certain of his emotions and psychic feelings. This book presents the romantic history of herbs and their use even today.

MENTAL ALCHEMY

by Ralph M. Lewis, F.R.C.

We can transmute our problems to workable solutions through *mental alchemy*. While this process is neither easy nor instantaneously effective, eventually the serious person will be rewarded. Certain aspects of our lives *can* be altered to make them more compatible with our goals.

Use this book to alter the direction of your life through proper thought and an understanding of practical mystical philosophy.

GLANDS—THE MIRROR OF SELF

by Onslow H. Wilson, Ph.D., F.R.C.

You need not continue to be bound by those glandular characteristics of your life which do not please you. These influences, through the findings of science and the mystical principles of nature, may be adjusted. The first essential is that of the old adage, "Know Yourself." Have revealed to you the facts about the endocrine glands—know where they are located in your body and what mental and physical functions they control. The control of the glands can mean the control of your life. These facts, scientifically correct, with their mystical interpretation, are presented in simple, nontechnical language, which everyone can enjoy and profit by reading.

THROUGH THE MIND'S EYE

by Ralph M. Lewis, F.R.C.

Truth is what is real to us. Knowledge, experience, is the material of which truth consists. But what is the *real, the true,* of what we know? With expanding consciousness and knowledge, truth changes. Truth therefore is ever in the *balance*—never the same. But in turning to important challenging subjects, the *Mind's Eye* can extract that which is the true and the real, for the *now*. The book, *Through The Mind's Eye,* calls to attention important topics for judgment by your mind's eye.

GREAT WOMEN INITIATES or The Feminine Mystic

by Hélène Bernard, F.R.C.

Throughout history, there have been women of exceptional courage and inspiration. Some, such as Joan of Arc, are well known; others have remained in relative obscurity—until now. In this book, Hélène Bernard examines from a Rosicrucian viewpoint the lives of thirteen great women mystics. Her research and insight have unveiled these unsung heroines who, even in the face of great adversity, have staunchly defended freedom of thought and the light of mysticism. Paperback.

THE UNIVERSE OF NUMBERS

From antiquity, the strangest of systems attempting to reveal the universe has been that of numbers. This book goes back to the mystical meaning and inherent virtue of numbers. It discusses the kabalistic writings contained in the *Sepher Yezirah,* and correlates the teachings of Pythagoras, Plato, Hermes Trismegistus, Philo, Plotinus, Boehme, Bacon, Fludd, and others who have explored this fascinating subject. *(Formerly published as "Number Systems and Correspondences.")* Paperback.

THE CONSCIENCE OF SCIENCE and Other Essays

by Walter J. Albersheim, Sc.D., F.R.C.

A remarkable collection of fifty-four essays by one of the most forthright writers in the field of science and mysticism. His frank and outspoken manner will challenge readers to look again to their own inner light, as it were, to cope with the ponderous advances in modern technology.

CARES THAT INFEST...

by Cecil A. Poole, F.R.C.

With a penetrating clarity, Cecil Poole presents us with the key to understanding our problems so that we may open wide the door and dismiss care from our lives. The author guides us on a search for *true value* so that, in the poet's words, "the night will be filled with music," as the *cares* "silently steal away."

INCREASE YOUR POWER OF CREATIVE THINKING IN EIGHT DAYS

by Ron Dalrymple, Ph.D., F.R.C.

Most people know they have the power within themselves to speak out and to do wonderful things, but they are constantly frustrated by not knowing how to get those great ideas out in the open. This new workbook contains simple but effective techniques designed to help you think more creatively. Its step-by-step program of exercises helps stimulate the flow of creative ideas by helping you develop a creative attitude and learn to think in creative patterns. Paperbound.

ORDER BOOKS FROM

ROSICRUCIAN SUPPLY BUREAU

ROSICRUCIAN PARK, SAN JOSE, CALIFORNIA 95191, U.S.A.

For a complete, illustrated catalogue and price list
of the books listed herein, please write to the
Rosicrucian Supply Bureau.